NEON TECHNIQUES AND HANDLING

NEON TECHNIQUES AND HANDLING

Handbook of Neon Sign and Cold Cathode Lighting

THIRD EDITION, 1977
Reprinted 1984

Written by
Samuel C. Miller

Edited by
Edwards R. Samuels

Published by
ST Publications
Cincinnati, Ohio

ISBN: 0-911380-41-8
Manufactured in the United States of America

Preface to the Third Edition

In the 25 years which have passed since the second edition of this book was written by Samual C. Miller, the electrical advertising and display field has changed tremendously. Various multi-colored plastics have been produced by the chemical industry. Special hot cathode lighting arrangements with ballasts instead of transformers have been applied to these materials. While neon has by no means been eliminated, in many applications it has been replaced by hot cathode. We have noticed a drop in the requirement for neon for many years in the electric sign field. However, in recent years there has been growth again, and neon is beginning to increase in its percentage share of the electrical advertising industry. The growth has been stimulated by the use of tubing in an increasing variety of applications, including contour lighting of plastic-faced letters, silhouette illumination, and artistic sign forms.

It was originally planned to include in this revised edition the necessary "how-to" of the plastic industry. This was not implemented because a study of the publications available has indicated many excellent publications which more than adequately provide the information required. The Rohm and Haas Company of Philadelphia, Pa., has produced a Plexiglas acrylic sheet design and fabrication data booklet which is a treasure trove of information. Another excellent publication is the fabrication manual published by the American Cyanamid Company of Sanford, Maine. The General Electric Company of Pittsfield, Mass., has provided a manual for the construction of signs using polycarbonate sheets going under the tradename of Lexan. Other publications are available by Eastman Chemical Company of Kingsport, Tenn., for the utilization of butyrate sheets known in the trade as Uvex and by the Filon Corp. of Hawthorne, Calif., utilizing fiberglass acrylic sheets known in the trade as Filon. There are also many other publications, and from time to time more and more will be added to the availability list.

The utilization of special light sources including hot-cathode and high-pressure lamps is available through the Large Lamp Dept. of the General Electric Company in Nela Park, Ohio. The General Electric Company Ballast Dept. of Danville, Ill., and the Universal Manufacturing Corp. of Paterson, N.J., as well as Jefferson Electric Co., Bellwood, Ill., and The France Mfg. Co., Westlake, Ohio, have provided excellent manuals dealing with the application of ballasts to hot-cathode plastic signs. They also have provided excellent information concerning the trouble-shooting and servicing of such signs.

The reader should be aware of the fact that there are many plastic fabricators in the industry who can provide anything the imagination creates in a prefabricated plastic element from letters through faces. Excellent information is provided in the current publication of the I.E.S. Lighting Handbook on lighting for advertising which deals with the quantitative requirements as well as the qualitative requirements of externally illuminated electrical advertising displays.

The author of this book, Samuel C. Miller, P.E., died in 1953. Sam Miller, the founder of Tube Light Company, was the man most responsible for the rapid rise of the neon industry. His engineering advice and compact neon plants, together with his many patented inventions, all were of great value in the rapid expansion of the luminous tube industry. It was indeed a severe loss to the undersigned to lose both a good friend and partner.

Appendix I, referred to in the second edition preface has been deleted, and readers are referred to the latest Underwriters' Laboratories requirements for construction of signs and also to all local requirements pertaining to the area in which the sign is to be erected. This varies from city to city and from town to town and should be carefully looked into before anyone attempts to contract to build an electrical display in any unknown territory.

This book has been carefully reviewed and, in some cases, minor changes have been made, keeping pace with the newer technical developments.

Edward R. Samuels

Moonachie, N. J.
January, 1976

Contents

PART ONE

Fundamentals Of Luminous Tube Manufacture

CHAPTER I

The Luminous Tube Industry

In 1910, Georges Claude demonstrated his first commercial luminous tube at the Grand Palais in Paris. In over 40 years which have elapsed, the luminous-tube industry has grown in the United States until, at present, its annual business is estimated at more than one hundred million dollars. From a single manufacturing plant in 1914, the number of manufacturers engaged in producing luminous tubes for signs, decoration, and lighting has increased to provide service in all parts of the country. The industry has grown rapidly in countries overseas, until today there is not a civilized country in the world that does not possess at least one plant engaged in the manufacturer of luminous tube. In many a South American city, the sign not only serves as an advertising medium but is used for lighting as well. If all the signs were turned out, in fact, the streets would be in complete darkness.

This world-wide business is done not only by large and well-established firms, but also by many hundreds of small prosperous plants which share in the total business. The largest plants may have invested many thousands of dollars in equipment, but the small and often successful sign craftsman has equipment which has cost him not more than a few thousand dollars.

The luminous-tube industry has a unique product to sell. In the early days of the business, there were many high-pressure promotion schemes, heavy competition, and wrangling over patent rights. But in recent years the industry has attained a high degree of stability. Continuous development of new materials and components for luminous tubes or lamps, in addition to refinements in the manufacturing process, have so improved the product and its adaptability to changing needs. In recent years, the introduction of fluorescent luminous tubes has not only increased the effectiveness of the advertising sign, but has, in addition, opened up a new market for interior illumination and decoration.

Why Is The Luminous Tube So Popular? Throughout the history of the luminous-tube enterprise, the public has shown a remarkable interest in luminous-tube signs. From the viewpoint of the man in the street, the first neon signs had the fascination of novelty, but surely no one can claim that luminous tubes are novel today. Long after the novelty wore off, the luminous-tube sign retained its sales effectiveness. There are many reasons why it has done so.

In the first place, the color of the light is easy on the eye but nevertheless extremely compelling. The great variety of colors available for sign tubing, due to the introduction of the fluorescent tube, has eye appeal which the old incandescent lamps could not offer.

A fact of little importance to the public but highly regarded by the advertiser is the high efficiency of luminous-tube signs. A pure neon tube radiates five times as much red light for a given amount of power as does an incandescent light. The fluorescent tubes used in most installations have approximately three times the efficiency for a given amount of power as do incandescent lamps. The brilliance and sparkle with which luminous tubes stand out on a rainy night are too readily apparent to require further comment.

Not the least important reason why luminous tubes are so well adapted to outdoor advertising is the flexibility which glass tubing provides for creating difficult designs. Many advertisers want to feature their trademarks. A trademark, reproduced faithfully in glass and in the proper colors, can be provided by the well-equipped sign manufacturer. Special figures, scrollwork, and animated designs are all well within the scope of the neon craftsman; in fact, there is almost no end to the possibilities afforded by tubular light. New applications, new forms, new mountings are being brought forward.

The original neon sign has been subject to the often-heard criticism that it was not a beautiful light, but this cannot be said of the present sign designed by a good craftsman; a well-designed sign adopts a combination of colors which is pleasing to the eye. This is why new signs and properly designed store fronts are now accepted by cities in which they are placed as positive contributions to the picture presented by the city as a whole, rather than as eyesores. Since its invention the luminous-tube sign has been accepted as an outdoor advertising medium of unquestioned value. But with the growing use of color balance and graceful forms, these signs are also beginning to be looked upon as a distinct credit to the concerns which install and operate them.

With the coming of fluorescent luminous tubing, the range of applications of tube lighting has been greatly extended. To the wide variety of colors possible with different gases and colored tubing, the use of fluorescent tubing has added a score of new pastel shades of high intrinsic brilliance. Of the greatest importance is the fact that fluorescent tubes have made possible for the first time a commercially practical form of artificial white light which can reproduce the effect of sunlight. The white light is, moreover, generated so efficiently that the fluorescent tube outranks all other forms of artificial light. This development has given rise to a wholly new field in the luminous-tube industry—interior lighting for illumination and decoration.

Two forms of interior fluorescent lighting have been introduced in recent years. The most widespread at present employs the low-voltage fluorescent lamp, or hot-cathode lamp, which is manufactured in standard sizes from 6 to 100 watts and operates with a low-voltage ballast. These hot-cathode lamps are usually manufactured by large companies on automatic machines and involve technical processes beyond the range of the luminous-tube sign shop.

In addition to the standard hot-cathode lamps, there has appeared a complete line of lamps known as Slimline lamps. These lamps are longer and larger in diameter than standard lamps and employ hot cathodes. Slimline lamps operate from ballasts at voltages up to 1000 volts.

The second type of fluorescent interior lamp is the cold-cathode tube. This is a conventional luminous tube and can be made by the average luminous-tube manufacturer. This type of lighting can be used either for direct or indirect lighting or for decorative use. For direct or indirect lighting many of the standard fluorescent white lamps can be used for best effect. However, for decorative lighting, any of the various combinations of colors can be used to give the desired effect and color scheme. Cold-cathode lighting can be operated either from special cold-cathode ballasts or from high-voltage transformers.

The advantage of this type of lighting is its inherent flexibility. The glass tubing may be straight or it may be formed in any shape required to fit the installation. Moreover, a given amount of illumination may be spread over a larger area with cold-cathode tubing than with the standard hot-cathode fluorescent lamps. This type of interior lighting should be part and parcel of the luminous-tube manufacturer's business, since it is in no essential way different from the production and installation of luminous tubes for advertising displays.

Finally, a primary advantage of tube lighting is the fact that it is a continuous form of light, extended over a line or surface. Hence it casts no sharp shadows and is particularly well suited to general interior illumination. Most other forms of artificial light, including the incandescent lamp, approximate point sources of light and hence cast shadows of the objects in the path of its rays.

The Background of the Industry. The first luminous tubes did not employ neon or any of the other rare gases. From 1893 to 1910, the so-called Moore tubes were prominent in the larger cities, but these were filled with nitrogen and carbon dioxide, two common gases. For reasons explained in the next chapter, these tubes had a very short life.

In 1910, Georges Claude introduced the first commercial neon sign, and from that time until the outbreak of the First World War he was busy bringing his discoveries to the point where they could be introduced on a large scale. Throughout that war, little of importance developed, although the fundamental Claude patent was granted Jan. 19, 1915.

Up to the time of the use of neon for signs, incandescent bulbs, glass letters, and floodlighted painted signs were used for outdoor advertising. But unless a great many bulbs of high wattage were used, the signs were weak both in visibility and penetrating power.

Trouble from broken bulbs and burned-out filaments made such signs costly to maintain. Between the time a bulb burned out and the time it was replaced, the sign became definitely unattractive.

By 1925, a number of concerns had been licensed under the Claude patents and had begun operations. Luminous tubes began rapidly to replace the incandescent-lamp signs, and the public interest in the neon displays grew steadily. Before 1930, the public demand for luminous-tube signs had grown to such proportions that the words "Claude neon" had become a household phrase. In fact, many people believed that the man who started the business was a man by the name of Neon. Within a short time, all the neon plants found themselves over-taxed with orders. This situation soon led to the formation of a host of new companies, some rivaling the Claude organizations. All the larger companies rode on the tidal wave of the business boom from 1925 to 1929.

The future of the industry seems assured. During the late '50s and the '60s the expanding popularity of fluorescent lamp illuminated plastic signs induced a number of electric sign companies to curtail, even close out, their neon production. However, the early '70s saw a trend toward increasing use of luminous tubing in silhouette effects and for illumination of plastic faced letters, symbols and other irregular shapes. This has sparked a promising resurgence of popularity for neon. It is a business which can return a fair profit on a very small initial investment, a business in which success depends largely on the ability of the sign or electrical craftsman to master the processes involved in tubing manufacture and to market the signs on a fair basis. It is becoming increasingly clear that the small plant, located in or near a medium-sized town, will supply the needs of that town over the competition offered by the more distant manufacturers. Transportation of luminous tubes is costly; and it has been shown clearly that the most economical way of selling and maintaining signs is to make them near the place of installation. It seems very probable, therefore, that each small town will have its local shop.

This book is designed to present the essential facts involved in the design, manufacture, installation, and maintenance of luminous-tube signs. As such, it is hoped that it may serve as a guide both for the man entering the business and also for the well-established craftsman who must occasionally solve unusual problems.

CHAPTER II

The Luminous Tube

The Luminous Tube: What It Is and How It Works. Fundamentally, the luminous tube is a very simple device. It consists of a vacuum-tight glass tube, fitted at each end with a metal terminal or electrode. Inside the tube is a small amount of rare gas. Connected to the two electrodes is a source of high-voltage electrical power. When the current is turned on, the tube glows with a steady piercing light.

The practical man can learn how to bend the glass, seal on the electrodes, pump out the air, fill the tube with neon or some other gas, seal it up, mount it, connect it to the transformer, and install the completed tube. He can do all these things without knowing how or why the tube works. But if his newly installed tube goes dead in two weeks or if his fluorescent tubing starts darkening to such a degree that it decreases visibility or illumination, and the customer demands satisfaction, the practical man can save himself a great deal of future trouble if he knows the simple facts about the glow discharge, what it is, and under what conditions it will deliver satisfactory service.

Sparks and Glow Discharges. The familiar glow of the neon sign is a first cousin of man's oldest electrical acquaintance, lightning. The red glow of neon and the blue flash of lightning are both electrical discharges in gas. In the neon sign, the gas is neon, a rare gas of great value. In lightning, the gas is the air, a mixture of oxygen, nitrogen, carbon dioxide, slight traces of hydrogen, and still slighter traces of the rare gases. In the luminous tube the pressure is very low—a partial vacuum, in fact—in lightning, it is the normal atmospheric pressure. But the neon glow and the lightning flash are electrically very much alike. They both illustrate the fact that, when electricity passes through a gas or a mixture of gases, light is given off.

If lightning and neon glow are so much the same, what makes their light so different? Just the differences given above, that is, differences in pressure and in the kind of gas used. The lightning spark is hot and blue-white in color. It consumes a tremendous amount of power. By using a particular gas and a reduced pressure in the luminous tube, it is possible to produce a steady glow, with very little power and very little heat. The combination of pleasing, penetrating light and high economy, with the flexibility provided by the glass tubing, makes the luminous tube a highly effective source of light for advertising purposes, lighting, and decoration. The large industry based on luminous tubes, as outlined in the previous chapter, has resulted from the practical application of this low-pressure electrical gas discharge.

What Happens Inside the Tube. To understand what happens inside a luminous tube, it must first be realized that the gas inside the tube consists of millions upon millions of particles called *molecules*. Under ordinary conditions, these molecules are electrically neutral; that is, they are neither positively nor negatively charged. To start the gas discharge, we must break these molecules up into electric charges. If we do this, a current can flow through the gas, and this current flow is accompanied by the desired light.

Each molecule is made up of negative *electrons* and positive *ions*. When the molecule is undisturbed, it has as many electrons as ions in it, and it is therefore electrically neutral. But if we remove a negative electron, the remainder of the molecule becomes positively charged; the result is the division of the gas into positive and negative charges.

The familiar law that electrical charges repel one another and that unlike charges attract each other leads to the fact that a free electron is strongly attracted to any nearby positive ion. This attraction leads to the rapid merging of the positive and negative electrons into one neutral molecule. During this process the remaining electrons, within the molecule, rearrange themselves and *the molecule gives off light*. This is the light we want to produce. Its color depends upon the kind of molecule, that is, upon the kind of gas.

Thus to produce a gas discharge we must continually remove electrons from neutral molecules and let them combine with positive ions to form other neutral molecules. The practical way of producing this ionization (as the electron-ion division is called) is by passing a current through the gas.

How the Current Produces the Glow. When voltage is applied to the electrodes, one electrode becomes, momentarily at least, positively charged, as shown in Figure 2.1. It attracts, therefore, any free negative electrons which may be present in the gas. *If the voltage is high enough,* the electrons will be attracted with tremendous force and will speed toward the positive electrode at a speed of thousands of miles per second. Before one of these electrons can get very far, however, it collides with a neutral gas molecule which lies in its path. It hits this molecule with such force that it liberates from it one or more (usually more than one) electrons. These electrons, once free, start off toward the positive electrode with great vigor. These electrons soon collide with other neutral molecules, liberating still more electrons, which go off to smash still other molecules. In an unbelievably short time after the voltage is applied to the tube, the whole body of gas is in a frenzy of motion, electrons being liberated from molecules, free electrons combining with positive ions, giving off light as they do so, and then being blasted apart again.

The voltage supplied by the luminous-tube transformer is

alternating; that is, it reverses itself so many times every second. In this country it reverses itself 60 times per second in a standard system. Because of this fact, each electrode takes its turn being the positive one, and as a result, the glow is distributed evenly over the whole tube. If direct current had been used, with one electrode remaining positive, the situation shown in (a) and (b) of Figure 2.2 would occur. This undesirable condition is avoided by the use of alternating current.

The Reason for High Voltage. Unless high voltage is used, the electrons will not be attracted strongly enough to start the

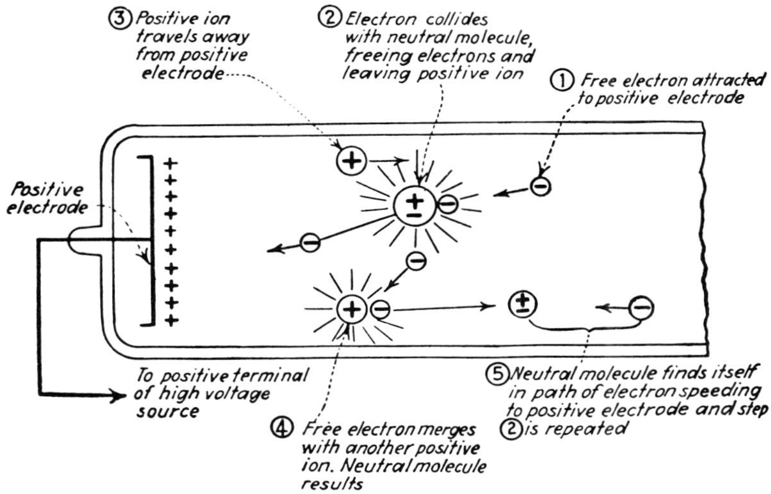

Figure 2.1. Action of electrons and positive ions in gas-filled tube.

ionization process, no current will flow, no light will appear, and the tube will not operate. Hence a high voltage must be used. Transformers are built with secondary voltages which vary from 2,000 to 15,000 volts for signs. In cold-cathode ballast systems, the secondary voltage is standardized at 750 or 900 volts.

Once the glow discharge has begun, however, less voltage is required to keep it operating. In fact, if the original high starting voltage was maintained, the current would be too great, and the gas would heat up excessively. The tube would consume a great deal of power, become very inefficient, and its life would be greatly decreased. Therefore, some means must be provided for lowering the voltage after the glow has started. A special type of transformer, made especially for luminous tubes and called by its manufacturers a *high-leakage reactance transformer,* performs this voltage reduction automatically. The method used is described in detail in a later chapter.

Operating Current. The electric current which passes through the luminous tube is an important quantity. If the current is too low, the tube will lack brilliance, if it is too high, the tube will overheat and have a very short life. The required operating current depends upon the diameter of tubing, the kind of electrodes used, the kind of gas and its pressure. Each of these factors must be correct so that the operating current is neither too small nor too large. The current of neon signs and cold-cathode tubing is measured in *milliamperes* (a milliampere is one-thousandth of an *ampere*) which is the standard electrical unit of

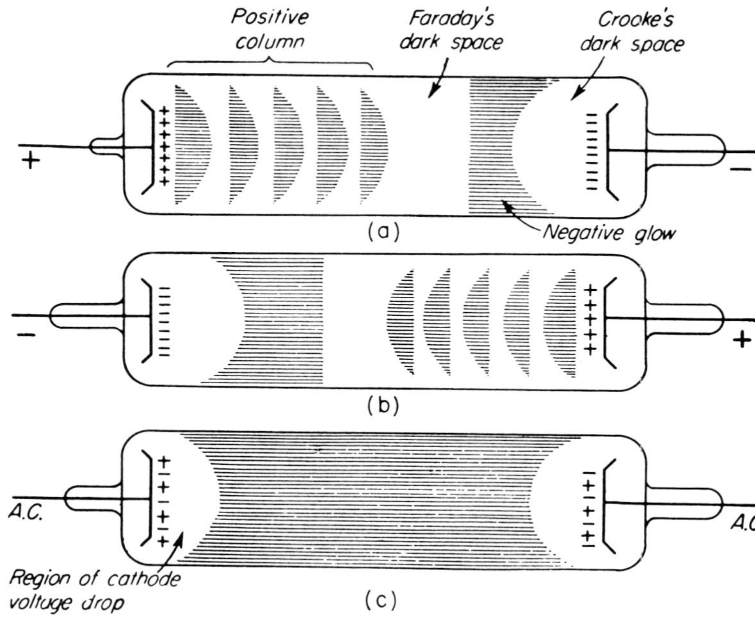

Figure 2.2. Schematic diagram of low pressure gas-filled tube. (a) and (b) show distribution of positive column when direct current is used. When alternating current is used, an even distribution as shown in (c) is the result.

current. Operating currents in standard signs run from 15 to 60 milliamperes. In cold-cathode lighting operating currents may run up to 120 milliamperes, and work is being done on lamps to carry 200 milliamperes.

The current through the tube is accompanied by the passage of free electrons from the gas into the electrodes. The heavier the current, the more electrons will flow to the electrodes in a given time. Each time that an electron hits the electrode, the energy with which it hits is transformed into heat, and as a result the electrode heats up. Hence, the heavier the current, the hotter the electrode becomes. If the current is excessive, the electrode will begin to disintegrate, or even to melt. Hence, electrodes must be designed to withstand considerable heat, although even with proper design they cannot stand up if the operating current is too high.

The Effect of Gas Pressure. We have already seen that a steady glow is possible only if the gas in the tube is at a reduced pressure. By "pressure" is meant the force with which the gas inside the tube presses against the glass wall of the tube. Pressure might be expressed in pounds per square inch, as air pressure is often expressed (normal air pressure, or "atmospheric pressure" is about 14.7 pounds per square inch). Usually, however, pressures in luminous tubes are expressed in terms of the height of the column of mercury which that pressure will support. At sea level, air pressure will support a column of mercury to about 760 millimeters (29.9 inches), and the pressure of the atmosphere is thus referred to as a pressure of 760 millimeters. When the tube is pumped out, the small amount of gas which still remains in the tube can exert only a small pressure, compared with that of the air. The gas in a luminous tube, for example, is usually at a pressure of less than 10 millimeters, or about 1/76 atmospheric pressure. Vacuum pressure used in luminous tubes, before filling with gas, often goes down to less than 0.001 millimeter.

The amount of current which will flow through the tube depends largely upon the pressure. To realize why this is true we must remember that in a gas at low pressure there are fewer molecules per cubic inch than in a gas at higher pressure. That is, in the low-pressure gas, the molecules are fewer in number and farther apart from one another. In the low-pressure gas, therefore, the free electron has a longer distance in which to get up speed before it hits a neutral molecule; as a consequence, when it hits, it hits much harder, more free electrons are liberated than would otherwise be free, and the whole action is much more intense. Hence, as the pressure is lowered, the current will increase. This is the general rule for pressures encountered in luminous tubes.

If, however, the pressure is reduced so low that a nearly perfect vacuum exists inside the tube, then there are so few molecules available that the current must decrease from sheer lack of electrons and ions, even though the electron speeds are very high. Consequently, it is found that, after a certain point is reached in lowering the pressure, the current begins to decrease again. The effect of pressure on current is shown graphically in Figure 2.3. In the range of pressures used in luminous tubes, between 3 and 20 millimeters, decreasing pressure means increasing current.

Summary. We now know that the current, and the resultant light in the tube, will increase if the pressure is lowered. We also know that excessive currents are extremely harmful to the tube, since they will injure its electrodes and shorten its life. The moral is plain. To make a successful tube, the voltage and pressure must be correct. They must be adapted to the gas or mixture used, and they must be maintained throughout the useful life of the tube.

The Effect of Diameter of the Tubing. Tubes for sign-lighting purposes are usually made with glass varying between 7 and 15 millimeters in outside diameter. These sizes are convenient to work with in practical applications. In many foreign countries, signs are made with larger tubing, up to 20 millimeters in diameter. In general, tubes larger than 15 millimeters in diameter are not used because the current available from standard sign transformers is limited to a maximum of 60 milliamperes. This amount of current can light effectively a 15-millimeter tube at standard pressure. If it were used with a larger tube, the light

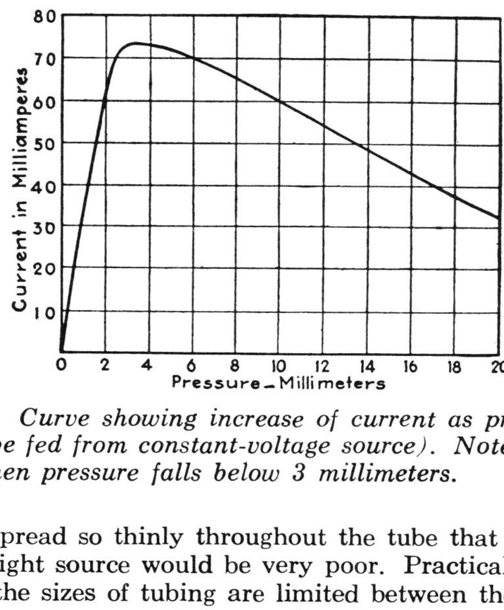

Figure 2.3. Curve showing increase of current as pressure is reduced (tube fed from constant-voltage source). Note decrease of current when pressure falls below 3 millimeters.

would be spread so thinly throughout the tube that its effectiveness as a light source would be very poor. Practically speaking, therefore, the sizes of tubing are limited between the two values stated above. In general, the smaller the tube used, the more brilliant the light. To obtain the maximum brilliance with the larger sizes of tubing, it is sometimes necessary to use slightly higher operating currents than those used with the smaller tubing. The smaller the diameter of the tubing, the higher is the resistance and a higher voltage is required to operate a foot of tubing. Figure 2.4 shows the relation between the necessary voltage and the diameter of the tube.

The foregoing remarks apply to luminous tubes of the non-fluorescent type. In fluorescent tubes, large-diameter tubes are often used, up to 25 millimeters outside diameter in the cold-cathode types, and up to 54 millimeters in the hot-cathode types. One reason for the use of the larger diameter of the fluorescent tubes is the fact that the light is very intense, and the surface brightness of the tube is reduced by spreading the light over the larger area of the larger diameter tube. This is particularly important in white tubes used for interior lighting, since too high a surface brightness may prove annoying to the eyes in interior installations.

The Gases Used for Luminous Tubes. No attempt will be made to go into details here concerning the gases used in luminous tubes, since they are fully described in the following chapter. In this introduction, only their electrical characteristics will be discussed, in relation to other electrical properties of the tubes. The rare gases, neon, argon, helium, xenon, and krypton, are ideally suited for use in signs, because the voltage per foot needed to produce a suitable brilliance of light is much less than that required for the more common gases, such as nitrogen and carbon dioxide, which also have been used in the past. Stated in other terms, the resistance of the rare gases is among the lowest of all the gases available for the purpose. The gases with the lowest resistance are not those which produce the most light, however. Argon, for example, has a very low resistance, lower than that of neon, but its light is comparatively weak. To obtain the advantages of low resistance and good light emission, gases are often combined, low-resistance gases being mixed with the good light

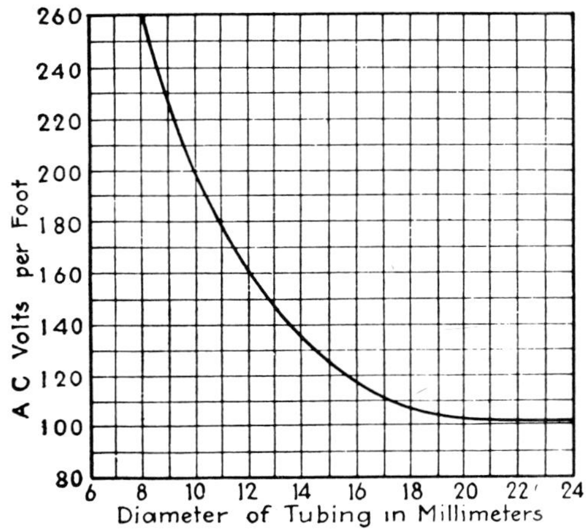

Figure 2.4. The voltage per foot required to operate tubing of various diameters filled with neon gas (no mercury).

givers to produce a compromise having the advantages of each type of gas.

The Use of Mercury. The most efficient and only practical way of producing blue light does not involve the use of a gas but of mercury vapor. Vapor differs from gases in that a vapor can exist only as it evaporates from a liquid, while a gas can exist of its own accord. To obtain mercury vapor, therefore, it is necessary to insert liquid mercury in the tube.

To set a curent flowing through mercury vapor is not so easy as through a gas, and for this reason a gas (usually argon or argon-neon mixture) is mixed with the vapor to aid the current flow. This extra gas also contributes somewhat to the light of the discharge, but its main function is that of a current carrier. The mercury vapor, like other vapors, is very much affected by changes in temperature; it will condense as the temperature is lowered. Thus, the light will become much more intense after the sign has had a chance to warm up.

In extremely cold weather, the sign may never come up to full brilliancy unless it is properly constructed. In installations subject to extreme cold, special starting gases containing helium are often used. The heat of the helium discharge maintains the mercury-vapor pressure. All these factors must be taken into account in designing a sign for the particular job that it is intended to fill. In general, the smaller the diameter of tubing used, the higher the resistance per foot and the hotter the tube will get, allowing the mercury to vaporize more readily. In cold weather, however, small-diameter mercury tubes are often found to fade out while larger tubes remain bright, because of the greater heat conduction of the smaller tube.

Chemical Effects Inside the Tube. The action of the luminous tube in producing light of the required brilliance and color depends wholly upon the electrical ionization of the gas within it. If the action could be restricted to a purely *electrical* one, we should not have to worry about the sign continuing to function properly. Unfortunately, there are many possible *chemical* actions which may start within the tube if special precautions are not taken. These chemical actions almost universally lower the efficiency of the tube. If any impurities, such as dirt, grease, or impure gas, are left in the tube after it is sealed off, these impurities, under the action of heat and electrical stress, become chemically active. As they combine with one another, they may blacken the glass, they may combine with the metal of the electrodes, or, if the heat is intense, they may liberate gas inside the tube. If this last possibility occurs, the unwanted gas will become ionized, and it will give off light, but not of the desired color and will become useless as a light source. This accounts for the fact that neon tubes often turn blue when they go bad. Many failures in tube making result from insufficient care in eliminating the impurities during pumping and filling. The gas inside the tube must be pure and the glass and electrodes must be thoroughly clean if undesirable chemical effects are to be avoided. Above all, the mercury must be the purest obtainable on the market.

Bombarding, an Essential Operation in Tube Making. Removing the chemical impurities is not so easy a task as may be supposed. Although the electrodes, the glass, and the rare gas may appear to be clean, they are actually harboring many impurities which cling to the surface and which cannot be removed by ordinary methods. The best way of getting rid of them is to heat the tube while it is being pumped out. The heat drives the gas and other impurities from the surface of the metal and glass in the form of some sort of gas or vapor. The pump removes these gases, and if the heating process is kept up long enough all of them are removed. The tube may then be filled with the rare gas and sealed off.

The most simple way of heating the tube and electrodes is by passing a current through the tube while it is still on the pump, when the air pressure has been reduced sufficiently to allow a heavy current to flow. This heavy current, usually much higher than the operating current, will pass through the low-pressure air in the tube, giving off much light and heat in the process. The light serves no purpose except to show that the bombardment, as it is called, is actually taking place. The heating has the desired effect of ridding the tube of its impurities. Bombardment is a most important and exacting process. A chapter (Chapter 9) has been devoted, therefore, to explaining what it is and how it should be carried out.

Electrodes and Tube Life. Sputtering. We have already learned these facts about electrodes: The electrode has the task of carrying current from the transformer wires to the rare gas. Since it is continually subjected to the bombarding of electrons and ions, it heats up, and it must be designed to withstand heat. Since the metal is hot, it is highly active chemically, and it may combine with gases or impurities within the tube. But by far the greatest difficulty with electrodes arises from what is known as *sputtering*. Sputtering occurs when the electrode, under the action of the heat to which it is subjected and the electrical forces which act upon it, flies to pieces bit by bit. The metal of the electrode gradually flies off and coats itself on the inside of the glass tube. This effect in itself would cause no harm, since the blackening caused by the metal deposit is confined to the ends of the tubes near the electrodes. Eventually, of course, the entire electrode would be consumed by the process, but since the action is very slow, the electrode will nevertheless last for almost normal life. However, sputtering is accompanied by a decrease of gas pressure in the tube. This loss of pressure eventually makes the tube inoperative.

The common explanation of sputtering action is as follows: As the metal electrode sputters to the glass walls, the metal particles trap the gas molecules and hold them between the metal coating and the glass wall. As a result, the pressure of the tube is greatly reduced. Finally the pressure is reduced so low that the sign loses brilliance, begins to flicker, and eventually goes dead. The gas in the tube is then said to have been "cleaned up" by the sputtering action. In practice, as the pressure falls off, the tube will operate at a very high temperature until the glass punctures or "sucks in" at its weakest point, usually near the electrode. The tube is then entirely useless, and it must be taken down, opened up, cleaned, repumped, bombarded, and refilled. In the early days, this sort of trouble was very common; in fact, the short life of tubes (due to sputtering) was one of the greatest hindrances to the commercial introduction of tube lighting.

Recently the theory of the "cleaning up" of gas due to sputtering, long held to be caused by trapping of the rare gas between the glass wall and the sputtered metal coating, has been challenged. Dr. Harvey Rentschler has proved that the gas, instead of being trapped against the wall of the tube, is absorbed by the metal of the electrode itself, after sputtering begins. But whether the gas is trapped against the wall of the tube or in the electrodes themselves, the result is the same. The gas pressure is reduced, the operating temperature rises, and the tube goes bad.

Sputter Control. The logical solution to the sputter problem is to build electrodes which can "take it," that is, electrodes which will not sputter under ordinary conditions. Some sputtering will always occur, and the life of the sign is thus always limited. But if the sputtering action is controlled, the life of the tube will be predictable, and maintenance guarantees and costs can be figured safely.

First, electrodes must have a large area exposed for the dissipation of heat. If an electrode runs cool, it will of course sputter much less violently than if it is hot. Secondly, the electrode must be made of the kind of metal which will best resist the sputtering action. Special metals and the common ones have all been tried; the best electrodes to use for each gas, each pressure, and each operating current are now known and have been more or less standardized by the industry.

The electrodes can be treated chemically to reduce sputtering before they are put in the tube; special patented processes are sometimes used for this purpose, and great virtues are claimed by each manufacturer for his particular "secret process." This treatment, while expensive, is sometimes absolutely necessary to ensure reasonable life for a sign. Treated electrodes are especially necessary for helium tubes. Special mechanical features are also incorporated in electrodes to reduce the tendency to sputter. All these features are treated at length in the section on electrodes in Chapter 3.

Mechanical Requirements of the Tubing. The maintenance of the proper gas pressure can be threatened by a more direct action than that of sputtering. If the glass envelope is not completely vacuum tight, air will leak in, and the sign will soon go dead. To assure complete airtightness, the glass should first be mechanically strong and thus not subject to cracks or other breakage. This requires great care in glass blowing, since, if the glass is cooled too quickly, it will be left in a very brittle and fragile state. Particularly for signs installed out of doors such strains within the glass will cause trouble, since the sign is then subjected to extremes in temperature.

The second requirement for a vacuum-tight tube is concerned with the lead-in wires which connect the transformer wires with the electrodes. Except for these lead wires, the whole wall of the tube is made of glass. The leads must necessarily be made of metal, to conduct the current. To make a good joint between glass and wire, the glass must "wet" the wire, that is, adhere to it firmly. Molten glass and wire do not readily cling to each other in this way unless the wire is copper or copper coated.

When wire and glass are heated, as when the wire is sealed on the glass or during the operation of the sign, both glass and wire expand. If the wire tends to expand more than the glass, it will press against the glass with great force. If this force does not crack the glass at once, it will subject it to a great strain which will crack it at the least provocation. It is important, therefore, that the glass and the wire expand at the same rate as they are heated. If they do so, no internal stress will occur, and the joint will be strong. In practice, a special alloy wire ("dumet" wire) is used for soft glass. This wire has a coefficient of expansion very nearly equal to that of the glass over the entire temperature range to which it will be exposed in manufacture and service. The seal will remain vacuum tight and serviceable if it is thus properly made. For pyrex glass, tungsten is used for the metal-to-glass seal.

Electrical Protection. Simple as a luminous tube is, it must be realized at all times that it is potentially a very dangerous piece of apparatus. When properly constructed and installed, it presents less danger than an ordinary lamp socket. But if proper precautions are not taken, the high voltage used is definitely dangerous. Since the current is limited to 30 or 60 milliamperes in most cases, the neon sign usually cannot produce a shock of sufficient force to kill a man. But the shock from a neon sign is powerful enough to overthrow a man, and the resulting fall may easily kill or cripple him. Furthermore, the possibility of the public coming in contact with the high voltage is far too great a danger to be allowed.

The installations for interior lighting using cold-cathode tubes, especially high-voltage systems, must be made and designed with special care, since the possibility of accidental contact or fire hazards are greater than in exterior sign displays. For such installations it is highly important that the manufacturer govern himself by the rules and regulations of the Underwriters' Laboratories and his local authorities and follow the procedures outlined by them. In this way, these installations can be made and installed with as little danger as the installations of any other electrical appliance.

The precautions which must be taken involve the thorough insulation of every metallic part in the high voltage circuit so that it may not come in contact with anything else. Simply putting the wires and electrodes out of reach of the prying fingers of children and innocent bystanders is not enough, because the wires may come in contact with metal moldings, showcase findings, etc., which will conduct the high voltage directly to the unsuspecting public. To avoid this, the industry has evolved, with the help of the Underwriters' Laboratories, a very complete set of insulating fittings and connecting wires which completely shield the high-voltage wiring not only from the public but from the weather as well. Protection from weather is essential in outdoor signs, of course.

The wires which lead from the transformer to the electrode leads are thoroughly coated with a high-resistance insulation capable of withstanding more than 15,000 volts. Special insulators of porcelain or pyrex glass are provided into which the entire end of the tube, with the electrode, fits. The connection of the high-voltage cable to the transformer is likewise protected by a porcelain bushing. The entire high-voltage circuit is thus completely enclosed with insulation from start to finish. When so protected, it is actually far less dangerous than the exposed lamp socket in the home.

The Operations Involved in Making a Luminous Tube. By way of summarizing the foregoing brief introduction of the fundamentals of luminous tubes, the following outline of the construction methods is presented.

A large workbench fitted with a fire-proof covering is required for laying out the patterns and for the actual work of glass bending. A supply of gas, and in some cases a gas booster, is required for raising its pressure. A supply of air likewise is necessary from a low-pressure air blower. A variety of gas burners, or "fires," connected to the gas and air supply, is used in heating the glass tubing to the melting point, so that it can be bent to the desired shape.

A complete pumping system is needed, capable of drawing a good vacuum in the glass tube; a supply of rare gas is used in filling the tube after it has been pumped. A high-powered bombarding transformer must be available for heating the tube and electrodes during the bombardment process. Special tools for working the glass and for sealing the glass when the sign is completed are necessary. A good vacuum gauge is essential for making the sign itself. A testing coil for testing for air leaks, etc., is a great convenience. A supply of metal, porcelain, and glass insulators is necessary for mounting the finished sign. A supply of tubing of various diameters of glass is needed. A supply of electrodes, rare gas, neon, No. 50 mixture, or argon is required. Clean mercury is vital, and some means of inserting it in the tube while the tube is on the pump must be provided. If all these parts, devices, and apparatus are available, work on the sign can begin.

Constructing a Luminous Tube Sign. The letters to appear on the sign are first laid out full size on drawing paper and then on asbestos paper. Charcoal or carbon paper is used to produce a layout pattern for each letter and for the glass connections between letters, borders, etc. The glass is then heated, bent to shape, spliced together, and the continuous tube, open at each end, is formed. A hole is made in the tube, and a small tube or *tubulation* is attached at a convenient point, for the purpose of connecting it to the pumps.

An electrode with its glass jacket is joined to each end of the tube. The tubulation is sealed to a corresponding size glass tube leading to the glass manifold. The glass stopcock attached to the blow hose is closed, and the stopcock to the pumps is opened. (For complete details, see Chapters 8, 9, and 10.)

After the pressure in the tube has been reduced slightly, as indicated by the vacuum gauge, the stopcock leading to the pump is turned off so that the pump is disconnected from the tube. If the tube is now thoroughly airtight, the low pressure will be maintained.

The bombarding transformer is then connected to the two electrodes of the tube. The main pump stopcock is opened, and the bombarder is thrown on. Just as soon as a discharge takes place, the main stopcock is closed. The bombarder is then turned on again, the tube glows, a heavy current passes, and considerable heat is generated. The glass gets so hot that it will scorch a piece of paper held against it. The electrodes eventually get red-hot. The bombarder is then turned off and disconnected. During this process, the main stopcock may have been turned on and off to produce the proper pressure within the tube. After a certain length of time, bombardment may take place with the stopcock wide open.

With the main stopcock open, the pump starts to reduce the pressure in the tube by exhausting impurities set free during heat-treatment. After pumping approximately five minutes (the time depends upon many conditions such as tube diameter and length of tubing) the main stopcock is then turned off, and another stopcock leading to the flask containing the rare gas is turned on slowly to admit a small amount of rare gas to the tube. The proper pressure is measured with a rare-gas pressure gauge, and, when it has been reached, the rare-gas stopcock is turned off. A tipping torch is applied to the tubulation leading to the pump, and the tube is "sealed off."

The tube is now complete except for the process of "aging," which is often necessary to bring it up to its proper brilliance. Aging is accomplished by connecting the completed tube to a transformer which will supply either the rated current or a current somewhat higher than rated. After a short period (not more than 15 minutes under ordinary conditions), the aging should be complete, and the tube will then operate at full brilliance on a transformer of the type which will be used in the installation. When the pumping system is in good working order and the tube is properly pumped before filling, no aging should be necessary.

Mounting the tube is a process which depends largely upon the particular installation. Usually, however, a background of metal is provided against which the tube glows and which serves to emphasize each letter. This background may be made, as it often is, in the form of a box or housing in which the transformer can be mounted. On the background are mounted stand-off brackets which seize the glass tube at several points and hold it firmly in place. At the electrodes, two recessed housings or bushings are provided for insulating the electrodes in the metal sign. From the electrodes or bushings, or from the housings to which the electrodes are connected, runs the high-tension cable which connects the electrode lead-in wires to each other or to the high-voltage transformer. The complete sign is shown in Figure 2.5. The whole sign, usually with the glass mounted and connected, is assembled and finished at the factory. It is then taken as a unit and installed complete at the location.

In some European countries it is the custom to make the letters of a sign individually in the shop and to assemble the sign during installation. However, this method is gradually losing favor because of the greater chance of breakage, excessive time on the customer's premises, etc.

The transformer is supplied with current through a low-voltage winding known as the primary winding, which is connected to

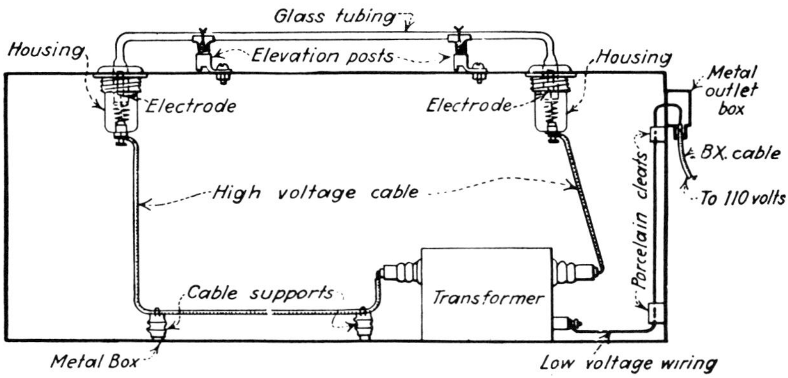

Figure 2.5. A schematic layout of a box-type neon sign, showing relative location of transformer, mounting parts, cable, tubing, etc.

the low-voltage circuit. The usual practice of using armored BX cable for house wiring is generally followed in the low-voltage connections. A time switch, for turning the sign on and off automatically at certain hours, may be connected in this low-voltage circuit. If a flasher, an automatic device for turning the sign (or its separate parts) on and off at regular intervals, is installed, this may be connected in the low-voltage circuit also. Flashers are often used in the high-voltage circuit.

Care is taken in hanging the sign so that it will not fall while in service, and it should be protected from the possibility of being broken by nearby objects.

Depending upon the care with which it has been designed and constructed, the sign, once installed, should give three to five years of continuous or intermittent service. At the end of that period, decomposition will have advanced to a degree where replacement is necessary.

The Maintenance Problem. The limited life of even the best constructed sign should not be forgotten by the sign maker who intends to make money in his business. He must know how long his signs can be expected to last under the various conditions to which they are likely to be subjected, and he must sell the sign in such a way that the customer understands the necessity for occasional replacement. Prices must be figured to take care of replacement, or if this is not done, the cost of replacement must be assumed by the customer. Not the easiest problem in the neon-sign business is that of figuring life guarantees and then living up to them. Costs will inevitably mount if the sign maker is continually servicing the signs that he has sold; therefore, he must know how to build a long-lasting sign. It is important to use the best possible material, since its cost is saved by reduction in maintenance expense.

CHAPTER III

Materials Used In Constructing Tubes

THE RARE GASES, ELECTRODES, GLASS TUBING

The Atmosphere. It was not until the middle of the seventeenth century that an English scientist by the name of Robert Boyle first attempted to analyze the true nature of the air. Since that time, we have come to know that the air we breath is made up of many different gases, some of them very plentiful and others extremely rare. Pure air, from which all the water, carbon dioxide, and dust have been removed, contains seven gases. The proportions of each are given in Table 3.1.

Table 3.1. Gases Found in the Atmosphere

Gas	Amount by Volume
Nitrogen	1 part in 1.281 (78.060 per cent)
Oxygen	1 part in 4.760 (21.008 per cent)
Argon	1 part in 106.8 (0.930 per cent)
Neon	1 part in 80,800 (0.001 per cent)
Helium	1 part in 245,300
Krypton	1 part in 20,000,000
Xenon	1 part in 70,000,000

The last five in the table—argon, helium, neon, krypton, and xenon—are called the *rare gases.* Just how rare xenon is may be realized from the fact that there is actually more gold dissolved in sea water than there is xenon in the air.

This interesting group of five gases forms a chemical family; that is, they display one definite chemical characteristic in common. All these gases are chemically *inert;* that is, they will not combine with any other substance; they are always found free, never in combination with other substances. Chemists, have never succeeded in making them combine either with each other or with any other chemical substance. This fact makes the rare gases ideally suited for use in luminous tubes. In the preceding chapter, we noted the injurious effects which may take place if the gas combines chemically with the electrodes or with impurities in the tube. The rare gases, because of their inert nature, cannot so combine.

Early Attempts to Use the Commoner Gases. It was natural that the more common gases, such as nitrogen and carbon dioxide, should be tried first in early attempts to produce successful luminous tubes. Although such gases are capable of producing a brilliant light, tubes in which they are used have a very short life. The main reason for this fact is the chemical activity of the gases used; the nitrogen or carbon dioxide soon combine with the electrodes. As this chemical combination goes on, the pressure becomes lower and lower, until eventually all the gas is "cleaned up," and the tube fails to function.

In 1893, D. McFarlan Moore began experimenting with luminous tubes using common gases. He conceived the idea of using an electromagnetic valve for admitting more carbon dioxide or nitrogen to the tube as the current through the sign changed owing to gas cleanup. This invention, while cumbersome, was instrumental in bringing the tube-sign industry to life, and Moore tubes, as they were called, were a familiar sight during the early 1900s. Nitrogen-filled tubes of large diameter were used to some extent in England in the past few years.

THE RARE GASES

The Discovery of the Rare Gases. The rare gases were discovered in the following order: Helium was found to exist in the sun by spectroscopic analysis, in 1868; it was discovered on the earth some 17 years later. In 1893, argon was discovered by Lord Rayleigh and Sir William Ramsay. In 1898, suspecting still other gases, Ramsay began a definite search for them and succeeded in isolating neon for the first time. A few days later, following the same method, he discovered krypton and xenon. The whole series of gases was thus revealed to science between 1880 and 1900.

The extreme rarity of the gases and the consequent expense of extracting them from the air made any commercial use an economic impossibility. It was not until 1907 that Georges Claude perfected a process for obtaining the rare gases cheaply. Working on processes which he and a German, Linde, had developed independently for the making of liquid air, Claude succeeded in laying the foundation of the luminous tube industry.

Commercial Methods of Extracting the Rare Gases. The process which Linde and Claude developed for the commercial extraction of the rare gases depends upon the liquefaction of air. Air, cooled by the usual refrigeration means, compressed by air compressors, and then allowed to expand suddenly through a small opening, can be made to lose heat. If the process is repeated, each time using the cool gas which results from the preceding expansion, the gas can be made successively cooler and cooler until eventually it becomes liquid. Liquid air, at a temperature of about —196°C (—371°F), contains liquid nitrogen, oxygen, and liquid argon. These three liquids have different boiling points. When the oxygen is allowed to boil, argon and nitrogen remain. The nitrogen is removed by chemical means, leaving pure argon. The same general process is applied for the separation of neon, helium, krypton, and xenon. Helium can be obtained much more cheaply from natural gas deposits, with which it is mixed in considerably greater amounts than in the air, but when so obtained it is not so pure as when extracted from the air.

Neon Gas. Neon, while not the commonest rare gas and while not discovered first, until recently was the most widely used gas in the luminous tube industry. It is colorless and odorless, slightly heavier than oxygen, twenty times as heavy as hydrogen (atomic weight), and five times as heavy as helium. It has an ionization potential (a measure of the voltage required to light it) lower than that of helium but higher than the other rare gases, argon, krypton, and xenon.

Since neon gas is not particularly well suited from the standpoint of either cost or electrical characteristics, we may well wonder why it has attained such universal popularity. The answers are its color, which at once caught the public fancy, and its high efficiency, which makes it one of the best sources of light for advertising purposes. The penetrating power of its red light, which shows up to best advantage in the poorest weather, has been utilized in beacons for aviation and marine service.

In an effort to establish the reason why a neon-filled clear-glass tube has this penetrating power, studies have been made of the light that it gives off. It has been found to be predominantly orange and red in color, but there are also small amounts of yellow and even green light. The relative amount of light of these different colors given off is shown in Figure 3.1. The peaks in the curve show that the greatest amount of light is given off in the red and orange regions.

Studies have also been made of the kind of light to which the eye is most sensitive. Figure 3.2 shows the relative visibility

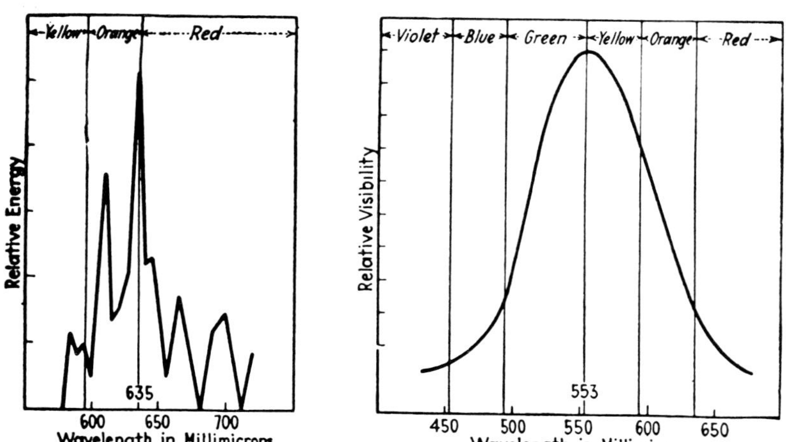

Figure 3.1 Relative light energy from a neon-filled tube. The light energy is confined primarily to the red and orange regions. Figure 3.2. Relative visibility of different colors to the eye. Maximum visibility occurs in green-yellow region.

of the various colors in the rainbow. It wll be seen that the maximum visibility is in the green-yellow region, while the orange-red region has much less effect. Why is it, then, that the neon glow has such a powerful effect? The answer is shown in Figure 3.3, on which is plotted the ease with which different colors are transmitted through the air. The relative transmission varies according to the kind of weather, as shown by the three curves. For rainy weather, the maximum transmission occurs at a wave

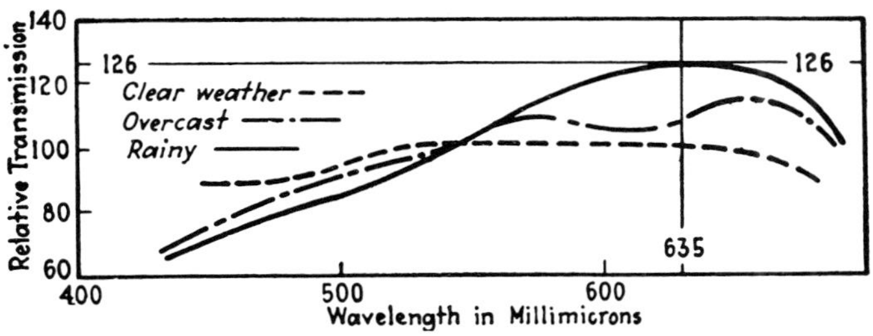

Figure 3.3. Relative transmission of different colors under various weather conditions. At 635 millimicrons, the wave length of neon light, transmission is highest in poorest weather.

length of 635 millimicrons, the wave length at which neon has its greatest output. This is the technical basis for the high penetrating power of neon and one of the most important reasons for its high popularity.

The Light Efficiency of Neon Gas. Since the neon tube was sold in competition, either directly or indirectly, with the incandescent light sign, the comparative cost of operating the two types of signs is a most important factor. Table 3.2 shows the light efficiency, in lumens of light per watt of power, for various gases and diameters of tubing. In comparison with these values, a standard 10-watt Mazda lamp gives 100 lumens of light, or approximately 10 lumens per watt. To get the same illumination value from a neon tube as that provided by a 150-watt lamp, 20 to 25 feet of 11-millimeter tubing would be required.

In sign work, however, the chief requirement is not illumination but visibility and, more important, readability. Since red rays have great penetrating power, they are the most important from the standpoint of visibility. A neon tube will provide 10 lumens per watt in the red range. The Mazda lamp transmits only 2 lumens per watt of the red. Hence, for the red ray the neon sign is five times as efficient as the Mazda lamp. This accounts for the generally accepted fact that, to produce equivalent penetrating effect, an incandescent lamp sign must be provided with two to three times as much power as the neon tube sign requires for the same effect. Also, if an incandescent sign were used with the same amount of light output as a neon sign, there would be a blurring effect because of the concentration of point sources of light in a small space, whereas the neon sign has its light distributed over a long length of tubing which tends to give better readability. Table 3.3 is a legibility chart, giving the distance in feet from the sign that various heights of letters are clearly legible.

Neon gas, as well as all other rare gases, is sold in glass containers, each containing 1 liter (61 cubic inches, or about the same volume as that contained in a quart bottle). In a new

Table 3.2. Light Output of Luminous Tubes

(Clear glass for red and blue; noviol glass for green)

Color	Rare gas used	Glass	Outside diameter, millimeters	Current, amperes	Lumens per foot	Watts per foot	Efficiency, lumens per watt
Red	Neon	Clear	11	0.025	70	5.7	12.2
			15	0.025	36	4.0	9.0
Blue*	No. 50	Clear	11	0.025	36	4.6	7.8
	or		15	0.025	18	3.8	4.7
	B-10						
Green*	No. 50	Noviol	11	0.025	20	4.6	4.3
	or		15	0.025	8	3.8	2.1
	B-10						

* Mercury used in blue and green tubes.

container, the gas is at atmospheric pressure, that is, the same as the air on the outside; so there is no pressure on the walls of the container.

Helium. Next to hydrogen, helium is the lightest gas known. It is found mixed with natural gas, is contained in certain minerals, and exists in the atmosphere. It derives its name from *helios,* the Greek word for "sun," and was so named because its existence was first shown in the sun by Janssen, who discovered a bright yellow line in the spectrum of sunlight. Later, helium was discovered on earth and identified as the same substance present in the sun.

In the luminous tube industry, helium was used to produce a yellow or whitish light. The gas has a high ionization potential compared with neon and the other rare gases; that is, it takes a higher voltage to produce the required brilliance of light. Because of their higher resistance, helium tubes take more power to operate than do neon-filled tubes and give off considerable heat. The gas must be used at low pressure (approximately 3 millimeters to give good brilliance). The resistance per linear foot of helium is about twice that of neon, and accordingly only half

Table 3.3. Legibility Chart

Height of letter	Kind of letter: Clear glass, neon-filled	Clear glass, No. 50 gas with mercury	Noviol glass, No. 50 gas with mercury
	Legibility, feet		
Inches			
2	65	50	42
3	100	75	65
4	150	115	98
6	200	150	130
8	350	260	227
9	400	300	260
10	450	340	290
12	525	395	340
15	630	490	410
18	750	560	490
24	1,000	800	650
30	1,250	940	815
Feet			
3	1,500	1,125	975
3½	1,750	1,310	1,135
4	2,000	1,500	1,300
4½	2,250	1,685	1,460
5	2,500	1,875	1,625
6	3,000	2,250	1,950
7	3,500	2,625	2,275
8	4,000	3,000	2,600
9	4,500	3,375	2,925
10	5,000	3,750	3,250
11	5,500	4,125	3,575
12	6,000	4,500	3,900

Note: The above information is only approximate for a person with average eyesight. The following variables change the above figures: current density, size of tubing, visibility (rain, fog, smoke), type of background used for glass letters, and the presence of other lights in the vicinity of the sign.

as many feet of helium tubing can be operated from a transformer as could be used if the tubing were neon filled.

The successful use of helium also depended almost entirely upon the electrodes used. Special electrodes had to be designed with chemical treatment intended to reduce sputtering. Practically all helium-tube electrodes have a special chemical coat which requires protection from moisture. The electrode is sold, therefore, in a moisture-sealed jacket. A more detailed account of these electrodes is given in the section on electrodes.

Since the advent of fluorescent tubing, the importance of helium in luminous tube technology has declined. Yellow and white colors, formerly obtainable only with helium, are now obtained from fluorescent tubes. The fact that the light output of a helium-filled tube is very low in comparison with a similar-color fluorescent tube and the severe footage limitations on the helium-tube transformer have practically eliminated the use of helium-filled tubes in the sign industry. Helium may sometimes be used, however, in combination with argon gas as a "carrier" gas in mercury-filled tubes for installation subject to extremely low temperatures.

Argon. Argon gas, from *argon,* "the inactive one," exists to the extent of one part in every 106 of the air we breathe. Suspected as early as 1796, by Cavendish, its existence was not definitely proved until 1893 by Ramsay, who named it "argon" because of its inert character. The gas is about twice as heavy as neon. In 1914, its commercial extraction was undertaken, not for the luminous sign trade but because it was found that incandescent lamps filled with argon had a much higher efficiency and longer life than vacuum lamps. About one-half of all the Mazda bulbs manufactured today contain argon.

Argon ionizes with a very dim lavender color. Its light output efficiency is not great; so it is rarely used by itself as a light source in gas-filled tubes. On the other hand, its extremely low ionization potential makes it an excellent conductor of electricity. It is used almost exclusively in the sign-tube art as the "carrier" gas in mercury-filled signs. This carrier action is necessary because mercury vapor within a sign tube will condense at low temperatures; when this happens, the tube will fail to start or else give a very faint light. Argon has the property of "supporting" the mercury vapor and keeping it active throughout a wide range of temperatures. Argon has a lower resistance than neon and permits current to pass more easily.

Recent experiments have shown that unmixed argon gas is the best carrier gas for mercury vapor, being superior in this respect to the argon gas mixtures commonly used. Because of this important factor and because argon has a low ionization potential, it has become the gas universally for cold-cathode installations where ballasts are employed.

Rare-gas Mixtures. Although many combinations of rare-gas mixtures can be employed to suit a definite need, it has been found that the best gas mixture for all practical purposes is the one known as No. 50 or B-10. This mixture contains approximately 80 to 90 per cent argon and 10 to 20 per cent neon as may be sold by different companies. It is used for almost every condition where a mercury tube is employed, either with clear or fluorescent glass, for luminous tube signs.

There have been other mixtures employed in the past. One was known as B-17 and contained 80 per cent neon and 20 per cent argon. This mixture, having a higher percentage of neon, has a higher resistance and has gradually been discarded. However, because of the higher neon content, it may still be used to give a slightly brighter tube where the number of feet of tubing per transformer is not an important factor. In this connection however, the brilliancy of this mixture, although great at first, diminishes after 200 or 300 hours' burning. Therefore, in general, the No. 50 mixture has been adopted for mercury-filled tubes.

The No. 20 mixture has also been used and may be used in some of the colder climates. It contains argon, neon, and a small percentage of helium which increases the resistance of the tubing and therefore keeps the tube at a higher temperature. Because of its higher resistance, it requires more transformers for equivalent tubing footage.

Krypton, although not used to any great extent now in the luminous tube field, may become an important gas because of its lower voltage drop and its lower ionization potential. The high price of krypton has been gradually decreased so that it now can be purchased at a price many times lower than ten years ago. Krypton is being used in new applications of hot-cathode fluorescent lamps.

Xenon is not used to any extent at the present time in the luminous tube industry.

Mercury Vapor. The first luminous tube ever constructed contained mercury. A simple evacuated tube containing mercury was found as early as 200 years ago to give a faint greenish-blue light when shaken. The static electricity generated by the friction of the mercury against the glass walls of the tube ionized the vapor, and light was produced. Today, mercury vapor is universally used for practically all fluorescent luminous tubes.

At ordinary temperatures, there is no such thing as mercury *gas;* ionization of mercury at room temperature can take place only if a voltage is applied to a body of mercury *vapor,* which cannot exist unless there is liquid mercury present. Hence, fluorescent tubes contain liquid mercury, which vaporizes under the proper conditions.

If mercury vapor alone were used, the operation of the tube would be very erratic, owing to the effect of temperature on the pressure of mercury vapor and also to the fact that the pressure of mercury vapor at ordinary room temperature is not sufficient to give a brilliant light. Hence, rare gas, either alone or in a mixture, is required to "support" the mercury vapor.

Mercury-filled tubes usually take some time before attaining full brilliancy; that is, the rare gas ionizes first, producing heat which vaporizes the mercury. Until the mercury has vaporized, depending upon the ambient or surrounding temperature, it will not contribute to the glow, and as a result a short time is required for the tubing to heat up and come to full brilliancy after it has been turned on.

Chemical Effects in Mercury Tubes. Mercury in tubes has a very great disadvantage which must be carefully considered in the design and construction of mercury or fluorescent tubes. Unlike the rare gases, it is very active chemically, especially when vaporized. It will combine with the electrodes or with any impurities in the tube unless the electrodes are especially clean and unless there is absolutely no foreign matter in the tube. The products formed when mercury combines with impurities are various shades of black or green in color, and they will darken the tube in short order.

A proper bombardment and the use of pure mercury, either triple-distilled or cathodic reagent, are essential to the successful manufacture of mercury tubing. Aging of the tube before the mercury is inserted is also recommended. These processes are all described in detail in later chapters.

Electrodes. Electrodes used for any type of luminous tube can be divided into two classes: the cold-cathode electrode and the hot-cathode electrode. In previous chapters, we have described the cold type. This is generally used for all sign tubing and cold-cathode lighting. The hot-cathode type of electrode is used in low-voltage fluorescent lamps. Since the sign or cold-cathode manufacturer does not make these low-voltage lamps, he seldom comes in direct contact with the hot-cathode electrode, but a knowledge of its operation is nevertheless helpful. Hot-cathode electrodes are described in Chapter 15.

Tubing with cold-type electrodes must be operated at higher voltages than tubing using the hot-cathode type of electrodes. Because of the convenience in the use of the cold-type electrode, the simplicity in degassing and processing, and the long life it gives the luminous tube, it has been adopted as the standard electrode for the sign and cold-cathode lighting manufacturer even though it requires higher operating voltage. This electrode has been the subject of a great deal of investigation by luminous tube engineers ever since the first rare-gas sign was installed, and as a result, a great deal has been learned about the design of the cold-type electrode.

Because of the difficulty in properly constructing electrodes in the neon sign shop, it is almost universal practice for the trade to buy the electrodes ready-made from supply houses. The electrodes supplied are of standard design and contain the following parts:

1. The electrode shell, which supplies the current to the gas in the tube.
2. The lead-in wires, which connect the electrode shell through the glass seal to the high-voltage cable and the transformer.
3. A glass jacket and seal, which surrounds the lead-in wires with a vacuum seal and which encloses the electrode shell in a cylinder of glass, open at one end. This jacket is joined or sealed to the end of the glass letter.
4. Some form of heat-insulating material inside the glass jacket which prevents the electrode shell from coming in contact with the glass. In some cases, sufficient spacing between shell and glass is used in place of other insulation.

It cannot be too strongly emphasized that success in building luminous tubes depends upon the use of properly designed electrodes. Hence, although the sign craftsman will not have to make them, he will profit if he knows how they are made.

Electrode Requirements. The electrode shell, being an electric terminal, must be made of metal. Since the shell is inside the tube, and since it is subject to heavy electrical and thermal stress, it must be made of a pure metal. If the metal contains impurities, these will be freed by heat during operation of the sign and will form a gas. This gas, mixed with the rare gas in the tube, will greatly detract from the sign's usefulness and efficiency. In fact, if the impurities in the electrode are excessive, they will cause the eventual breakdown of the tube. There are many metals which can be used for the electrode shell, namely, iron, nickel, and copper. In general, most electrodes today are made from iron of high purity.

The second requirement for a good electrode is sufficient area for heat dissipation and for the purpose of carrying the tube current without sputtering. This requirement was the basis of the fundamental patent issued to Georges Claude. Since the electrode gets hot during bombardment, it must be able to dissipate this heat without melting. It must also be large enough in surface area to withstand the electric forces which act on the electrode in operation and which may cause it to disintegrate slowly or to sputter.

The degree of sputtering depends to a large extent upon what is known as the *cathode-voltage drop,* or simply, *cathode fall.* When the electrode is negative, as it is many times a second during operation, the positive ions are attracted to it. These ions, being much heavier than electrons, bombard the surface of the electrode with great force and cause it to wear away slowly. The force with which these positive ions act depends upon the drop in voltage at the electrode. The higher the cathode fall, the greater will be the bombardment and sputtering.

To keep the cathode drop low, two methods are available. The first and simplest is to make the electrode large in surface area. The second, more complicated but very widely used, is to coat the electrode chemically with a coating which reduces the cathode fall by liberating free electrons.

Another device for reducing sputtering of the electrodes is a collar of insulating material which covers the external edges of the electrode, at the opening of the shell. This confines the sputtering to the inside of the electrode shell and does not allow any particles to escape from the shell and contaminate the tube. The tendency to sputter is correspondingly reduced.

A third requirement for electrodes used in mercury tubing is the prevention of chemical reaction. When used with the chemically active mercury of the fluorescent tube, the electrodes may be injured by chemical action since chemically active mercury combines with metal electrodes especially if they have not been thoroughly degassed. Table 3.4 illustrates the voltage drop of various types of standard electrode shells, with and without coating.

Lead-in Wires. The lead-in wires must be able to carry the sign current without heating, and they must have the same coefficient of expansion as the glass in which they are enclosed. To meet these requirements, a special wire has been developed for use with the soft glass universally used in the sign or cold-cathode industry which is a copper-clad wire known as *dumet wire*. This is an alloy of nickel and iron (about 40 per cent nickel) which is covered with a thin jacket or sheath of copper, the copper cross section being as much as one-fifth the cross section of the wire. The wire is treated with a borax solution and borated to give a liberal coating of oxide which allows for better adherence or "wetting" to the glass. This wire has ample current-carrying capacity; it has a coefficient of expansion closely matching that of lead glass, and the borated copper coat makes it adhere to the molten glass which is pressed around it.

Table 3.4. Voltage Drop of Neon-sign and Cold-cathode Electrodes

(Iron shells, nickel plated; operated at rated current on a mercury tube)

Electrode size, millimeters	Voltage drop inside the shell	
	Without coating	With coating
15 (shell: 3/8 inch diameter × 1 5/16 inches long)	200	120
19 (shell: 1/2 inch diameter × 1 5/8 inches long)	170	90

The seal between the wire and the glass must be completely vacuum tight. The use of dumet wire practically ensures this, but care must be taken in manufacturing the seal to allow the glass to cool slowly; otherwise it will contain internal strains which will crack the seal. The wire diameter used for lead-in wires varies from 16 to 30 mils (0.016 to 0.030 inch).

Other lead-in wires for soft glass have been developed, such as the Allegheny No. 55, Sealmet, and Sylvania No. 4 metal.

The majority of electrodes are made with copper-clad leads attached to the seal and brought out in one piece for making the electric connection on the outside. However, in some cases, where electrodes are used in corrosive atmospheres, as in factories or seaside locations, a three-piece weld is employed with the electrode. The three-piece weld consists of an iron or nickel piece attached to the electrode shell, a small piece of copper-clad wire for sealing into the glass, and another nickel lead extending out for the terminal. The nickel lead extending as a terminal or tie wire can be either a solid or stranded wire.

For hard or pyrex-glass seals, tungsten is used for the lead-in wire. There also has been developed a special metal known as

Kovar which makes a perfect seal between the hard or pyrex type of glass and the metal.

Electrode Insulation. The electrode shell is kept from coming in contact with the inside of the glass jacket by the use of an insulating material. Mica is almost universally employed for this purpose, although other forms have been used such as lava collars, glass beads, solid glass jackets, usually made of pyrex and completely surrounding the metal shell and fiberglass. The purpose of this insulating material is to keep the hot electrode from touching the glass jacket, either during operation or during the bombarding process. During bombardment, if the shell touches the glass, it will crack the glass. During operation of the sign, if the electrode shell accidentally touches the glass, there is a possibility of cracking at that point because of the difference in temperature of the inside and outside of the glass jacket.

Some manufacturers rely on very strong and stiff lead-in wires to keep the shell centered within the glass jacket. This method is good if the shell can be made to keep its centered position at all times.

The glass jacket which forms part of the envelope of the completed sign must be carefully designed also. It must be of convenient cross section to seal onto the glass tubing. It must not be too thick, or else the high heat to which it is subjected during bombardment will crack it; and it must not be too thin, or it will be mechanically weak. Also, the electrode must be perfectly annealed so that the strains are eliminated from the glass, particularly from the press or seal.

How Electrode Shells Are Made. Most electrode shells in general use today are made by automatic eyelet-drawing machines and are furnished in sizes which have been adopted as a standard in the industry. These shells may vary from shell units ⅛ inch in diameter and ½ inch in length to those having a ½-inch diameter and a 2-inch length. There are many different sizes and lengths in between these two limits.

The shells are open on one end and closed on the other. They are furnished in two types, the cup and the pinch type. In the cup-type shell, which is entirely closed on one end, the lead-in wire or wires are welded to the closed end. The pinch-type shell is closed on one end, with the exception of a very small opening through which the lead-in wires are inserted. The metal is then clamped down between two punches, squeezing the lead-in wires between the metal and making a solid connection. The various types of shells are shown in Figure 3.4.

The wall thickness of the shells must be sufficient to withstand heating effects during bombardment and during operation. The wall thickness of the standard shells varies from 0.010 to 0.018 inch, depending on the size of the shell used. At the present time, electrode shells are being made to carry up to 150 milliamperes of current and research is being conducted for cold types of electrode shells to carry higher currents.

Processing the Shell. It has been found that, if the shells are specially processed during manufacture, their usefulness and uniformity will be definitely improved. Preheating in a hydrogen furnace serves to eliminate many of the impurities and gases which have been absorbed or occluded in the metal or in the nickel plating and to prevent the shells from rusting.

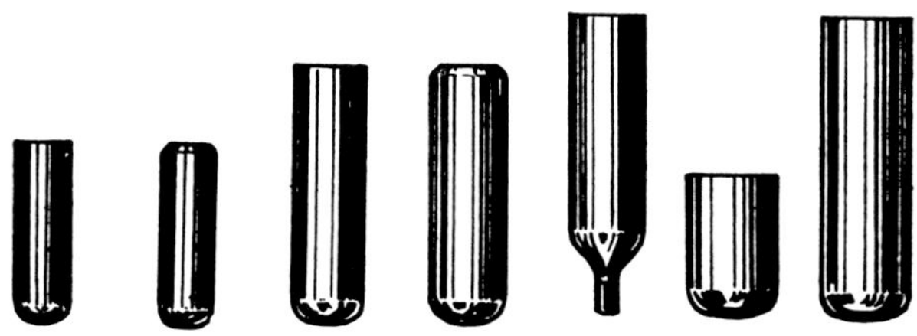

TYPE	9C OR 10C	9CI OR 10J	12C OR 12	12CI OR 12J	12P	15SH	15
SIZE SHELL	1/4" D x 11/16	1/4" D x 11/16	5/16" D x 1 1/8	5/16" D x 1 1/8	5/16" D x 1 1/4	3/8" D x 5/8	3/8" D x 1 5/16

TYPE	15CI OR 15J	15P	19SH	19M	19CI OR 19MJ	19L	19LP
SIZE SHELL	3/8" D x 1 5/16	3/8" D x 1 7/16	1/2" D x 7/8	1/2" D x 1 5/16	1/2" D x 1 5/16	1/2" D x 1 5/8	1/2" D x 1 3/4

Figure 3.4. Standard cup- and pinch-type electrode shells.

Processing to Reduce Cathode Fall. When the electrode shells are not coated, a high resistance develops around the electrode, and the positive-ion bombardment gives rise to excessive sputtering. To meet these conditions, the metal shells are coated with an electron-emitting substance. The coated electrode is particularly useful to the manufacturers of cold-cathode tubing who use ballasts. Ballasts are designed with secondary voltages of 750, 800, or 900 volts. The voltage drop of the electrode becomes an important factor in the starting and operating of a tube and a ballast.

The coating most commonly used is barium strontium carbonate, which is reduced to barium strontium oxide during the heating of the tube in manufacture. The oxide coating is a good source of free electrons, which reduce the voltage drop at the surface of the electrode and hence reduce the tendency to sputter. Coated electrode shells particularly require a great deal of care and a high degree of heat for processing during the pumping procedure. It is very important that the coating be broken down into oxides and all the impurities released during this process. Failure to bombard the electrode shell properly may cause trouble later on. Therefore it is important to bring the shell to the proper temperature during the bombarding process.

If untreated electrode shells are used, the voltage drop is considerably higher, although the processing of a tube is simplified, but untreated electrode shells may cause greater sputtering and undue blackening.

When used with mercury-vapor tubes, the coated electrode must be carefully "formed," that is, the carbonates mixture must be thoroughly reduced to oxide before the mercury is admitted to the tube; otherwise the mercury will combine with the carbonates and cause a greenish darkening which means dirty fluorescent tubing. This will be noticed immediately after the tubing is pumped and should not be mistaken for the black darkening which takes place because of impurities or sputtering.

Types of Seals. The most common method of making a seal for an electrode is the pinch external seal. The same glass used for the electrode jacket is melted down at the end of the glass. The glass is pressed between clamps over the lead-in wires and forms a vacuum-tight seal, as used in the many different types of electrodes illustrated in Figure 3.5. To form a pinch-type seal properly, the thickness of glass must be such that it is neither too heavy nor too thin. The seal must not be too cold; otherwise

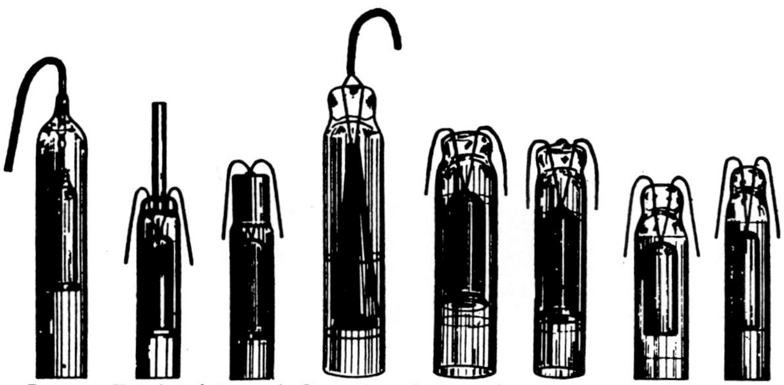

Figure 3.5. Standard types of electrodes.

it will cause an open seal or a "leaker," but neither should the wires be burnt. Also, the glass toward the bottom of the seal must be blown out to prevent thickness of wall. A properly made pinch seal should show very little strain when viewed through a polariscope. Pinch-seal electrodes can be furnished with or without tubulation.

Although not commonly used, the flare or external ring-seal electrode has been used in the past and may become more commonly used in the future, especially for cold-cathode lamps. In the past, this type of seal was supplied complete with the glass jacket, as illustrated in Figure 3.6. In the future, only a stem and electrode shell welded to the lead wires may be furnished

RING SEAL ELECTRODE

Figure 3.6. In this type the shell is first attached to a glass stem and then sealed to the electrode glass tube.

and the sealing-in will have to be done by the sign or cold-cathode manufacturer himself. Stems for ring seals can be furnished with or without tubulation attached to the stem.

The use of a ring seal in a straight cold-cathode tube will give a minimum of dark space at each end because the fluorescent glass can be brought right to the end and sealed to the stem.

The metal-to-glass seals show great promise for future use. These metals can be formed in either an elongated cup with a beveled edge, similar to the present electrode shell, or in a flat metal disk. The elongated unit can act as both the electrode and the unit to be sealed directly into the glass. With the disk type, a standard electrode shell can be spot-welded to the inside of the disk. The electrode shell then acts as the electrode, and the disk, with the proper coefficient of expansion characteristics, is sealed to the glass. A great deal of care must be taken in making a metal-to-glass seal. The metal itself should be oxidized, preferably in a wet-hydrogen furnace. Metals usually used for making a metal-to-soft-glass seal are Allegheny Sealmet and Telemet and Sylvania No. 4 metal. For a hard-glass seal such as metal to pyrex, Kovar metal is most satisfactory.

With hard glass such as pyrex or nonex, a tungsten lead-in wire must be used. Because of the stiff, inflexible nature of the tungsten wire, copper pigtails are often fastened to the tungsten to avoid breakage in connecting and handling the electrodes. The tungsten lead-in wire is first ground and polished to avoid microscopic cracks within the body of the metal itself. This is to prevent slow leaks which sometimes may develop in tungsten-pyrex seals because of these microscopic cracks in the wire. The tungsten lead with the electrode attached is usually prepared by sealing over it a bead of pyrex, nonex, or other hard glass. The beaded unit is then sealed into the electrode jacket. It must be understood that pyrex, nonex, or any other so-called hard glass will not seal to the standard lead, lime, or soda glasses known as soft glass. Figure 3.7 shows soft-glass and pyrex-glass electrode mounts.

Pyrex is not now used for luminous tubes except in certain European countries. It is used for special applications; for example, pyrex glass is used for glass condensation pumps because the glass will withstand greater changes of temperature. It is also used universally for laboratory and shop apparatus.

Insulation Materials. The first method which was used for insulating shells in luminous tubes was suggested by Claude in his original work, that is, rings of glass beads fastened around the outside of the electrode shell. However, more modern forms have included mica disks, ceramic and porcelain collars, and mica sheathing.

The method used almost universally today employs a mica shield. The mica shield consists of a thin split piece of mica about 0.001 inch thick, rolled on a tube and put into the electrode from the open end after the electrode has been fabricated. The mica shell is thin and flexible and presses very closely against the inner wall of the glass tube. In some cases, the mica shield extends at least $\frac{1}{8}$ inch beyond each end of the shell and gives it full insulation protection. In other cases, the mica sheath is small, extending $\frac{1}{8}$ inch beyond the front end and only about halfway back along the length of the shell. In this case, the manufacturer uses the rigidity of the support wire to keep the shell centered and insulates only the front end.

The material used for insulating the electrode shell must, of course, contain no impurities which may be liberated inside the luminous tube after it has been pumped. For this reason, furnace

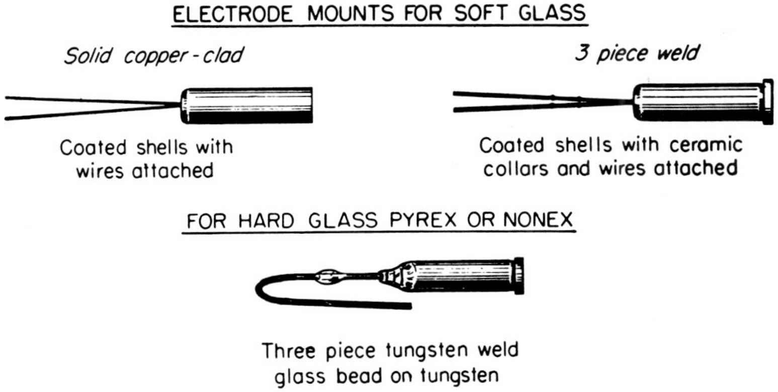

Figure 3.7. Various types of electrode mounts.

heating is desirable before these parts are attached to the electrode.

Hot-cathode Electrodes. Hot-cathode electrodes can be divided into two classes: (1) electrodes which are maintained at the proper temperature by a continuous flow of current, and (2) electrodes which are brought up to temperature initially, after which the heating power is cut off and the heat of the electrode is maintained by the arc discharge through the tube. The first type is not generally used for luminous tubes or lighting because it requires a separate power source for heating the electrodes and keeping them at a heated temperature. This type of electrode is more generally used for vacuum tubes, cathode-ray tubes, etc., in the radio industry.

The second type is used for the standard hot-cathode fluorescent lamp. The filament is composed of tungsten wire of proper diameter, coiled over mandrels and coated with an electron-emitting substance. The coiled electrode is attached to a two-wire stem and sealed on each end of the fluorescent tube. When the switch is thrown on in the case of a standard hot-cathode fluorescent lamp, the filaments at both ends are connected through a starter to allow them to attain incandescence. As soon as this is accomplished, the starter circuit is broken and an inductive surge through the ballast causes an arc through the lamp. When the lamp lights, the arc keeps the points on the filaments at each end at incandescence and liberates sufficient electrons to cause a low voltage drop. The low-voltage cathode drop, approximately 14 to 18 volts depending on electrode size, cathode emission, and type of gas and pressure used, allows such a lamp to operate from the line voltage.

Glass Tubing. Glass for tube signs or cold-cathode lighting must have many well-defined properties. It must be strong, and it must be permanent, even when exposed to variations of temperature and weather and to processing during manufacture. It must melt at a convenient temperature so that it may be worked easily in the gas fires ordinarily available. It must withstand working strains imposed upon it such as taking the glass directly from the fire and placing it upon asbestos patterns. It must not contain excessive impurities, and it must be uniform both in diameter and thickness of wall.

The following excerpt from a Corning Glass Works *Bulletin* describes the essential characteristics of glass:

It might be said that glass nature is very much like human nature. Treat it in the way it wants to be treated and it yields to your will. Abuse it and a more unpredictable, uncompromising material would be hard to find. For one to know right from wrong in glass working it is necessary to understand thoroughly the basic character of the material. This knowledge plus some reasoning will help to avoid trouble or to diagnose it properly when it does occur.

Glass should be thought of not as one material, but as two. Under certain conditions it is a solid and at other times it is a plastic material which is capable of flow and can be stretched, bent, or otherwise permanently reshaped. The change from one state to the other is brought about by a change of temperature, as is well known. There is no specific temperature at which the transition from one state to the other occurs. As glass becomes hotter it becomes increasingly fluid. Conversely, as the glass cools, it becomes stiffer. Thus we see that it is incorrect to refer to the "melting point" or the "freezing point" of glass because no such points exist. It is possible, however, to visualize a temperature range which may be called the annealing range below which glass is essentially a solid with practically no ability to flow and above which it is a plastic and can flow with relative ease. It must be remembered that, in the temperature regions within or just above the annealing range, the ability of glass to flow is not discernible to the eyes if an attempt is made to bend or distort the glass. In this range the glass appears solid, but actually it can flow very slightly, and this is important to annealing as will be seen later. The above facts are true of all glasses and vary only in degree as glass composition changes.

Annealing is a process of heat-treatment which either eliminates or prevents the formation of permanent cooling stresses in the glass. These stresses may cause the glass to break if they are allowed to remain. This breakage may occur as the glass cools, or it may occur some time after a sign is put in service. A rope if pulled taut enough will break because of tensile stress; likewise a stick bent across the knee breaks for exactly the same reason. Glass also breaks for exactly the same reason. Stress is the result of opposing forces.

It is a generally known fundamental physical fact that most materials increase in volume or size when they are heated and that they decrease in size when cooled. This is also true of glass. For every degree change in temperature there is a certain amount of change of size. The amount of change in length of each unit of length for each degree change in temperature is known as coefficient of expansion. For different glasses this value is different. If two layers of a single piece of glass were at different temperatures, we would expect stress to result from the different amounts of contraction or expansion if both of these layers were in the temperature region where they could be considered solid and unyielding. If one layer were in the plastic range, then no stress would result because the plastic layer would yield to the change in size of the solid layer.

If, as a piece of glass cools from the plastic state, it enters and passes through the annealing range (transition from plastic to solid) too rapidly, a nonuniform temperature within the piece results, and this produces permanent stress. If this same piece of glass were cooled through the annealing range so slowly that the temperature within the piece was always uniform at any instant, then the glass would reach the solid state without stress. This process is called annealing. It is very important that all glass should be annealed if it has been heated to or above the annealing range.

There are many different types of glass used throughout the world for signs and cold-cathode lighting, but in general these can be classified into two groups, soft glass and hard glass. The soft glass group can be further divided into *(a)* lead and *(b)* lime or soda glass. In the hard glass group, pyrex has usually been used. During the past ten years, most of the manufacturers throughout the world have adopted the American standard by using lead glass for luminous signs. In some European countries, lime or soda glass and pyrex are still used for luminous tube signs. For cold-cathode manufacture, both lead and lime-tubes are used for lighting purposes. As a rule, lead glass is used where tubing has to be bent or shaped. Lime-glass tubes for lighting are used only where straight lengths are employed, the only heat applied being that to seal on the electrodes to the ends of the tube.

Lime glasses are made from about 70 per cent silica sand. The remaining ingredients are mainly lime and soda ash. Since these materials are inexpensive and relatively easy to melt and shape, lime glasses cost less than the other types.

Lead glass contains about 60 per cent silica sand and 30 per cent lead oxide; the remainder is mainly potash and soda ash. Lead glass has a high electrical resistivity necessary for certain types of radio, radar, and television tubes and is used also for nearly all the luminous tube signs.

Although glass tubing has been drawn by hand for many years and is still drawn by hand for special purposes, most standard lead-glass tubing is now made by automatic machines.

Automatic Production of Glass Tubing. The glass is first melted in a large continuous tank in which the basic ingredients are mixed, and the molten glass is drawn from the machine on long rollers. In one method the molten glass flows by gravity over and around a hollow rotating mandrel through which the air flows under pressure. As the molten glass flows around and over the mandrel, it is formed into tubing and drawn several hundred feet along the rolls. At the end of the rolls, it is cut into standard lengths by automatic cutting machines. Glass tubing is also produced automatically by a machine which operates on a vertical updraw plan.

In the past, many types of glass of various compositions and colors have been used for luminous tubing. Recently, these have been standardized so that they fit practically all needs for present-day practice. Most of the colored glasses used before the war have been eliminated because of the advent of fluorescent tubing which allows a greater variety of colors that can be produced in solid-colored glasses.

At the present time, four types of glass are in common use, namely, clear lead glass, soft-red or ruby glass, noviol glass, and soft-blue glass. Table 3.5 gives the Corning code description and various physical properties of these glasses. Code 001 is used for the sign and lighting industry. Code 012 may be submitted but is used more frequently in the radio-tube industry. The table refers to softening point, annealing point, and strain point. These occur at arbitrarily selected viscosity values which denote the degree of fluidity of the glass. The softening points is a point of value to the manufacturer in his production control. It gives an indication of the temperature at which many glasses will begin to distort noticeably or flow when heated. The temperature range between the annealing point and strain point indicated in the table is referred to as the annealing range and is the temperature range wherein proper cooling is necessary to avoid a permanent strain or stress in the glass. The table indicates that Code 3530 noviol has an annealing point approximately 80 to 85°C higher than either Code 001 or 012. This is the hardest of the soft glasses listed. It should also be noticed that the coefficient of expansion is very nearly the same in all the glasses. Therefore, no trouble should be experienced in sealing any of the glasses to each other if the annealing range is adjusted to accommodate the ranges of the two glasses involved.

The physical properties of other types of glass that may be used are also indicated in Table 3.6. These are a clear soda-lime

Table 3.5. Physical Properties of Neon-sign Tubing*

Corning code	Description	Softening point, °C	Annealing point, °C	Strain point, °C	Expansion coefficient
001	Clear lead	626	428	397	91
012	Clear lead	630	433	400	89
2475	Ruby	690	480	447	93
3530	Noviol	693	510	478	91
5320	Blue	675	505	475	91

Table 3.6. Physical Properties of Other Types of Soft and Hard Glasses*

Corning code	Description	Softening point, °C	Annealing point, °C	Strain point, °C	Expansion coefficient
0080	Clear soda lime for lamp bulbs	696	510	478	92
7050	Borosilicate (pyrex) for series sealing	703	496	461	46
7052	Borosilicate (pyrex) for Kovar sealing	708	475	435	46
7720	Borosilicate (pyrex) for electrical use	755	518	484	36
7910 and 7911	96 per cent silica for ultraviolet transmission	1500	910	820	8

* Corning Glass Works.

glass for cold-cathode bulbs, borosilicate or pyrex glasses for apparatus and other uses where a hard glass is necessary, and a silica or quartz type of glass for ultraviolet transmission.

Standardization of Glass Tubing. Over a long period of time a great deal of work has been done in standardizing glass tubing, not only as to tolerance in composition of the glass itself, but also in the physical dimensions. These include tolerances in variation of diameter, wall thickness, and the permissble amount of "bow" in a length of the glass tubing. Standardization also includes tolerances of color matching. Colored tubing is held within color limits so that a uniform color results and tubing fabricated a number of years earlier can be matched with tubing of current manufacture. Table 3.7 gives the diameter, approximate number of 4-foot sticks, feet and meters per pound, and meters per kilogram of clear lead-glass neon sign tubing.

How Clear and Colored Glasses Are Used. Clear glass can be used by itself to produce either a red-orange luminous tube when filled with neon alone, or it can be used with a rare gas and mercury to produce a blue tube. It can also be used with helium gas, although this is not the current practice, to give a pink-white tube. With fluorescent coating clear glass is now used almost universally to produce the many variations of colors which can be obtained with fluorescent powders.

Noviol glass, when filled with mercury and a rare gas, gives a yellow-green color. This is because noviol tubing filters the violet and blue color emitted by mercury discharge and allows the transmission of the yellow and green. Although the matching

Table 3.7. Data for Lead-glass and Colored Tubing*

(Standard 4-foot lengths)

Diameter, millimeters	Approximate sticks per pound	Approximate feet per pound	Approximate meters per pound	Approximate meters per kilogram
8	4½	18	5.5	12.0
9	4	16	4.9	11.0
10	3½	14	4.3	9.5
11	3¼	13	4.0	9.0
12	3	12	3.6	8.0
13	2¾	10½	3.2	7.0
15	2¼	9	2.7	6.0
18	2	7¼	2.2	5.0
20	1¾	6¼	1.9	4.0
22	1½	5¾	1.7	3.7
25	1¼	5	1.5	3.3

* The table gives the approximate data for both clear- and colored-glass tubing used for neon signs. Clear-glass tubing may vary from 0.050 to 0.060 inch in wall thickness. Colored-glass tubing, noviol, blue, and ruby may vary from 0.040 to 0.070 inch in wall thickness. Standard tubes may vary from 46 to 48 feet in length.

of color of noviol tubing may not always appear uniform in different batches of tubing, still the resulting color is the same because of the filtering characteristics. Noviol tubing is also used in combination with fluorescent powders, as will be explained later, to produce a gold color. Noviol can be used with green fluorescent powder to produce a dark-green tube.

Soft blue tubing filters everything but blue light. Blue tubing can be used with fluorescent powder to produce a dark blue, although this is not commonly the practice.

Soft red or ruby tubing filters all but the red light and a small amount of the red-orange. Ruby tubing can be used with fluorescent powder to produce a dark red, although this is not common practice.

It must be understood that colored glasses such as noviol, ruby, and blue are filters and merely change the color of the light transmitted from within the tube by absorbing unwanted colors as shown in the light transmission curves in Figure 3.8 *(a)*, *(b)*, and *(c)*.

A Blok-Out or black tubing made from clear glass is now available for sign use. This tubing is used in small sections for joining letters together. As it retains its dark color permanently, it has some advantage over the usual method of painting or blocking out that portion of tubing between letters. This tubing is made from regular clear sign tubing by a special treatment which produces black lead near the surface, but actually within the glass itself. The use of black or block-out tubing requires more time and labor because of the extra splices which have to be made. However, it offers a permanently black, block-out portion and does away with the need of periodically repainting peeled painted surfaces.

Fluorescent Tubing and Materials. Although fluorescent tubing has been in use for a good many years, it is only within the past ten years that it has been developed to a very high degree for luminous tube signs. At the present time, the method used for producing fluorescent tubing is to coat the inside of clear-glass tubing with a very fine layer of fluorescent powder. Previous to the use of coated-glass tubing, a number of solid-glass fluorescent tubes were produced. One type of tubing made in this country obtained a series of different colors from fluorescent materials dissolved in the glass itself. Another type, produced by a German concern, was made in two layers, a clear-glass layer on the outside of the tubing and a layer of fluorescent glass on the inside surface. Neither of these fluorescent glasses was satisfactory because their fluorescent brilliancy was not very much greater than that obtained from the clear or colored glasses in use. Furthermore, the breakage due to strain was very high. These solid fluorescent glasses required a great deal of skill, care, and annealing which were beyond the scope of the average shop.

With the advent of the coated fluorescent glass, an important step forward in sign and lighting methods was taken. The glass used is the standard glass commonly used for clear-glass signs with the addition of the proper chemical coating on the inside surface. The color of the light produced depends upon the chemical used. The basic fluorescent phosphors most commonly used are indicated in Table 3.8.

Basic powders can be used in conjunction with each other to produce various color combinations or preblends. Preblends most commonly used are Daylight 6500, 3500, 4500, Warm White, and Soft White. The numbers such as 6500, 3500, and 4500 indicate the color temperature in degrees Kelvin.

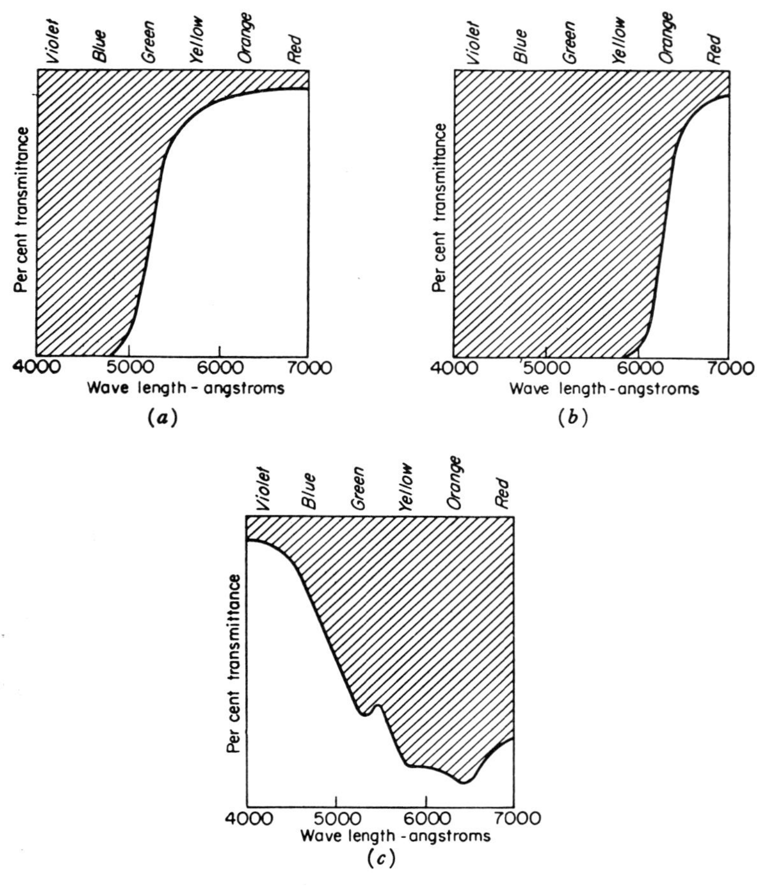

Figure 3.8. Typical transmission curves of colored glasses. (a) Noviol glass No. 3530. (b) Ruby glass No. 2475. (c) Blue glass No. 5320. (Corning Glass Works.)

The shaded area of each spectrum curve represents light absorbed by the glass; the clear area represents transmitted light. Ruby glass, for example, absorbs almost the entire spectrum, and transmits only in the red region.

In connection with the use of fluorescent powders, there are factors other than the chemical composition of the powders which may affect the ultimate color produced. These are: gas and the pressure used, diameter of the tubing, and the operating current through the tube. In commercial practice, the blended powders are usually marked for the diameter of the tubing for which they have been prepared in order to obtain the necessary color.

Table 3.8. Basic Fluorescent Phosphors Most Commonly Used

Basic Powder	General Color
Calcium tungstate	Blue
Magnesium tungstate	Blue-white
Calcium silicate	Pink
Zinc silicate	Green
Calcium halophosphate*	Yellow-pink-white
Cadmium silicate	Yellow-pink
Cadmium borate	Pink
Calcium cerium phosphate†	"Black light"
Barium alumina silicate†	"Black light"

* Color depends on percentage of activating agent.
† Provides ultraviolet light to illuminate fluorescent-painted surfaces.

When a tube is coated with any of the basic powders or blends, and then processed, pumped, and filled with a rare gas, the fluorescent particles become excited because they are subjected to ultraviolet light produced by the mercury discharge through the tube. The mercury ultraviolet light has an energy peak at a wave length which corresponds to the excitation value of the fluorescent phosphors, or 2537 A. This means that the maximum amount of energy conversion will take place, and it is for this reason that the present types of fluorescent phosphors have been developed. It

Table 3.9. Fluorescent Phosphors
(Transmission in angstroms)

Phosphor	General color	Exciting range* (ultra-violet)	Sensi-tivity peak	Emitted range (visible)	Emitted peak
Calcium tungstate	Blue	2200-3000	2720	3800-7000	4400
Magnesium tungstate	Blue-white	2200-3200	2850	3800-7200	4800
Zinc silicate	Green	2200-2960	2537	4500-6200	5250
Calcium halophosphate	Yellow-white	2200-3000	2537	4500-7200	5950
Cadmium silicate	Yellow-pink	2200-3200	2400	4300-7200	5950
Cadmium borate	Pink	2200-3600	2500	4000-7200	6150

*2200 is lower limit of measurements.

is noticed (Table 3.9) that the excitation value is in the ultraviolet range and consequently not visible. All fluorescent powders used in our industry have the property of transforming this invisible ultraviolet energy into high-level visible energy as indicated in Table 3.9. In theory, they operate like a radio receiving set that receives a signal of radio frequency not audible to the human ear and transforms this inaudible energy into the audible signal. The fluorescent powder used must be of the highest quality to obtain maximum conversion. The powder must be subject to constant manufacturing inspection, careful blending to give the proper color at all times, and the coating thickness inside the tubing must be uniform. If a tube is coated too thinly, a great deal of the light from the positive column will go through the pores of the coating and the coating will not be fully utilized. If the coating is too thick, it may cause an excessive amount of absorption and also may cause the powder to break away at the bends of the letters.

Although the maximum amount of energy, and therefore light, is obtained from fluorescent tubes using mercury, it has also been the practice to obtain certain colors without the use of mercury. The excitation range of the fluorescent powders is wide enough to produce a high level of visible energy even when excited by a gas-discharge tube filled with neon gas only.

Fluorescent tubing and any of the powders used for coating the tubing are particularly susceptible to contamination from impurities. For this reason, both the tubing and the chemicals must be kept free from dust and water vapor. Tubing should be kept tightly corked, and chemicals should be kept in a dry and clean location. Care must be exercised in handling the finished tube to prevent dislodging the powders from the glass.

The fluorescent coating may be applied to the inside of the glass in either one of two ways. The more simple and less expensive method involves coating the inside of the glass tube with a liquid mixture of fluorescent powder, methyl, butyl, or ethyl alcohol and a trace of resinous binder. The liquid can be either forced into the tube by air pressure or sucked into the tube by a vacuum pump. Figure 3.9 shows a standard filling rack using air pressure. When the tube is filled with the liquid and the liquid is allowed to run down again, a deposit of fluorescent coating and alcohol is left on the inside of the glass wall. Air is then blown into the tube, and the alcohol quickly evaporates, leaving the deposit of powder alone. This method of applying fluorescent powder is used in many cases by the small sign shops which buy a fluorescent mix and make their own tubing as necessary. However, owing to the fact that a quick, low-boiling-point drying agent is employed, coating by this method is not very satisfactory. Rapid evaporation of the alcohol may cause the coating to become heavy, or too much alcohol may cause a thin coating. Quick drying may cause streaks, and the coating may not have a uniform thickness. Also, the adherence of the chemical to the glass walls may not be very substantial. Therefore, the powder may break away when subject to slight jars or shocks or to the sudden rush of air into the tube caused by breaking the tubulation while a tube is being processed on the pumps. As a quick and inexpensive method of making tubing, this method may be employed. Figure 3.10 shows a vacuum method of coating fluorescent tubes.

The second method of applying fluorescent powder to a tube is to use a combination of powder and a standard lacquer consisting of acetates, nitrocellulose, and binder. Such a liquid has a high boiling point, and when a tube is coated with this solution, the drying takes places at a much slower rate so that a more uniform coating is obtained. Because of the organic content of some of the chemicals employed in such a mixture, these tubes must be subjected to heating in an oven to a temperature of 850°F to

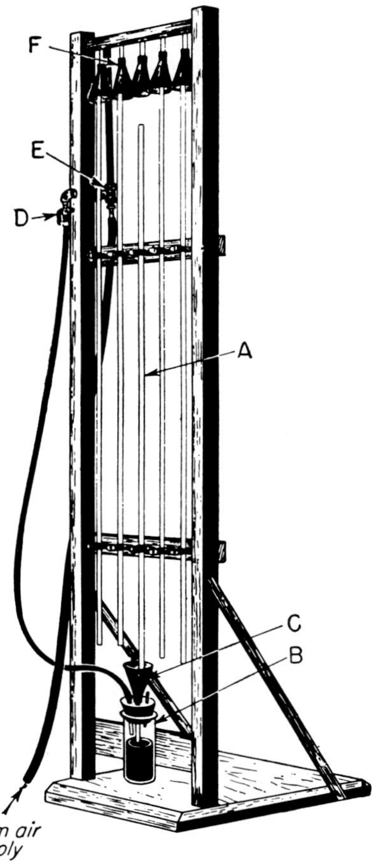

Figure 3.9. Glass coating rack using fluorescent cold mix. Forced-air operated. A. Tubes to be coated. B. Glass jar with fluorescent mix. C. Rubber funnel. D. Manually operated valve for regulating air pressure to force liquid mix up into glass tube. E. Valve for regulating flow of air through upper funnels (F) for drying tubes.

remove all organic material. At this heat, sintering takes place between the tubing and the glass which gives a better adherence between the powder and the inner glass wall. Tubing made by this method has a much more uniform distribution of powder, uniform thickness, and stable performance. This is the generally accepted method used for high quality signs and is the method used for lighting wherein the light output is of paramount importance.

A third method used in many European countries is radically different. The glass tubing is first bent into shape according to a pattern. Then the inside of the bent tubing is coated with a mixture of phosphoric acid, camphor, and alcohol. This leaves a sticky surface on the inside of the tubing. The fluorescent powder, in a dry state, is then dusted into the tube. The improper use of phosphoric acid in this connection may cause undue darkening of the tube and consequent loss of efficiency.

As can be readily seen, the variety of colors which can be obtained by the use of fluorescent-coated tubing is unlimited. However, it has been necessary to standardize on a range of colors most suitable for neon sign and lighting applications. Most

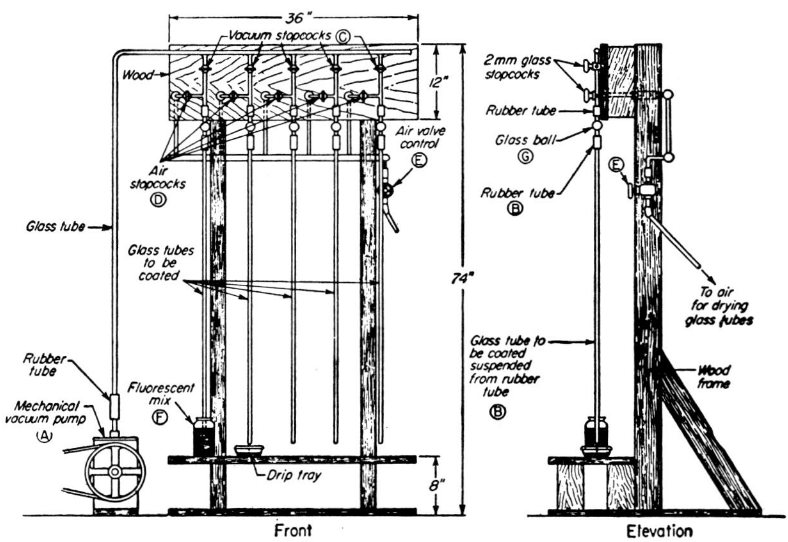

Figure 3.10. Glass coating rack using fluorescent mix. Vacuum-pump operated. Each glass tube is individually coated. It is attached to the system by means of rubber tube B and glass ball G. When stopcock C is opened slowly, the vacuum in the manifold produced by pump A will cause the fluorescent mix F in the jar to rise slowly in the tube. When the liquid has reached the top of the tube, the vacuum stopcock is closed and air stopcock D is opened. The liquid will fall back to the jar, and the air flow regulated by valve E will evaporate the binder and leave a fluorescent coating on the inside of the tube.

manufacturers have standardized their color range with each other. Table 3.10 indicates the range of colors which is normally available from most of the manufacturers of coated fluorescent tubing. There may be some variations and additions since a great many manufacturers have special colors which they have adapted for special purposes. The table indicates most of the standard colors available.

The sign or lighting manufacturer must acquaint himself with all the available colors and decide upon those most advantageous for the sign or lighting installation. For sign purposes it is necessary not only to have a brilliant and beautifully designed sign of pleasing color but also to be sure that the message or text is readable. Many signmen use very bright colors for letters and then place them so close together that, although brilliant, they produce a blurring effect and are not readable. For lighting

Table 3.10. Standard Fluorescent-coated Tubing for Luminous Signs

Fluorescent tubing designation	Kind of glass used	Filling gas	Color produced
Green	Clear	No. 50 and mercury	Standard green for brilliant sign work
Dark green	Noviol	No. 50 and mercury	Deep subdued green
Amber or neogold	Clear	Neon	A nonfading gold very popular for signs
Yellow or noviol gold	Noviol	No. 50 and mercury	A bright gold popular for sign work
Blue	Clear	No. 50 and mercury	A subdued blue for sign work
Dark blue	Soft blue	No. 50 and mercury	A deep subdued blue
Rose or neopink	Clear	Neon	A nonfading rose color very popular for sign work
Light pink or pink-white	Clear	No. 50 and mercury	A reddish-pink color used for sign work and also for adding color in interior illumination
Warm White	Clear	No. 50 and mercury	A peach-white
Daylight White	Clear	No. 50 and mercury	A cold white for illumination
3500 White	Clear	No. 50 and mercury	A warm-toned white for illumination
4500 White	Clear	No. 50 and mercury	A cool white for illumination
Dark red	Soft red	No. 50 and mercury	A bright ruby red

purposes the whites are most generally used. These are: 3500, 4500, and 6500 Daylight. In some applications Soft White is also used for lighting. In many cases of cold-cathode lighting, especially in bars, halls, or other places where a subdued and soft light is desired, the various colors such as neon-pink, blue, or whites in combination with each other can be used to get the proper color and light.

Fluorescent Mix. Fluorescent mix is supplied by manufacturers of fluorescent powders to those who wish to coat their own tubing. It consists of basic fluorescent powders combined with the necessary alcohol and a minute quantity of resinous binder, mixed together in the proper proportion and milled to the required consistency. When the powder is ball-milled to the proper consistency, it has approximately a 400 mesh, much finer than that of the basic powders which have a 200 to 300 mesh. Fluorescent mix can also be obtained in the same colors indicated in Table 3.10 for fluorescent-coated tubing. It is supplied in 500 or 1000-cc glass jars.

CHAPTER IV

Electrical Equipment

Electrical Accessories for Tube Signs. The luminous tube requires an electrical power supply, and since the supply must be of high voltage, particular care must be taken with the insulation and mounting of the various electrical accessories. Under the general heading of electrical equipment are the following:

1. Insulators and mounting parts, including elevation or tube posts for mounting the tube proper, housings or rings for shielding the electrodes in the metal sign, and cable supports for mounting the high-voltage cable.
2. The high-voltage cable, which connects the electrode terminals to each other or to the luminous tube transformer.
3. The luminous tube transformer, which steps up the line voltage (usually 115 volts in the United States) to the high voltage required for the tubes.
4. The low-line voltage wiring, including BX cable, line-voltage switches, and connections to junction boxes.
5. Special switching and animating devices, such as flashers, spellers, time clocks, and photoelectric relays.
6. Rotary converters for obtaining a supply of alternating current where only direct current is available from the power lines.
7. Power-factor correctors, for improving the power factor of the luminous tube transformer. Transformers may be obtained with or without power-factor correction. If a transformer is used without power-factor correction, condensers can be added to correct the power factor. Usually it is necessary to correct the power factor from 85 to 90 per cent or better.
8. Radio-interference eliminators, for reducing the radio noise caused by sign installations.

If all these types of equipment are understood by the sign engineer, he will have little trouble in handling the electrical problems which arise in tube making and installations.

One important piece of equipment used in the manufacture of signs will not be treated in this chapter. It is the bombarding transformer and its regulator, devices not used as accessories to the completed sign but of fundamental importance in making the tubes. This equipment is treated in the chapter on bombardment.

Insulators and Mounting Parts. The elevation post, or "tube

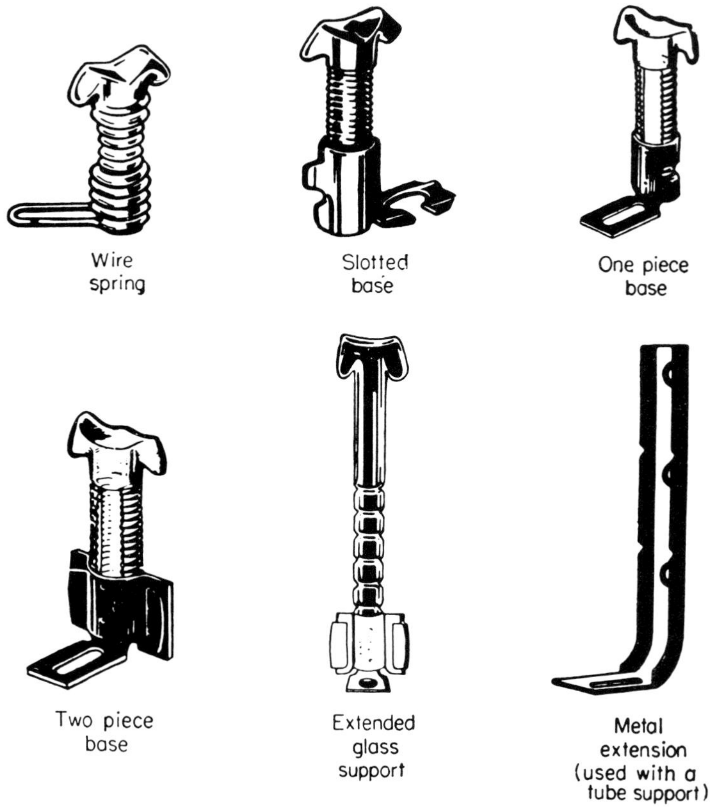

Figure 4.1. Various types of tube supports and their mountings.

support," is an insulator used in mounting the tube. At the bottom of the post is a metal bracket for fastening the post to the body of the sign. The total height of the post, including the insulator and metal fastening base, varies from about $1\frac{1}{4}$ to $3\frac{1}{2}$ inches according to its use. At the top of the insulator, provision is made for gripping the glass tube securely. The standard glass type contains a semicircular recess in which the tube is held by a thin wire. This wire is wrapped around the tube and the flanges on the post.

Two or more posts are used to grip each separate letter or section of the tube. To keep the tube parallel to the base on which it is mounted, the posts must be of the same height. Under certain mounting conditions, it is necessary to have one post shorter than the others in order to keep the tube parallel to its background. For this reason, elevation posts are usually made adjustable in height. Special extension posts are used in window-border signs.

Electrode Housings. Housings are made from dry-process or wet-process porcelain or pyrex glass. The housing fits into the sign box and is mounted rigidly in place. The tube is so placed that, when it is held by the elevation posts, the electrode jackets fit into the housing. The phosphor-bronze spring inside the housing bears against either the electrode lead-in wires or a metal cap at the electrode end, making contact directly with them. The

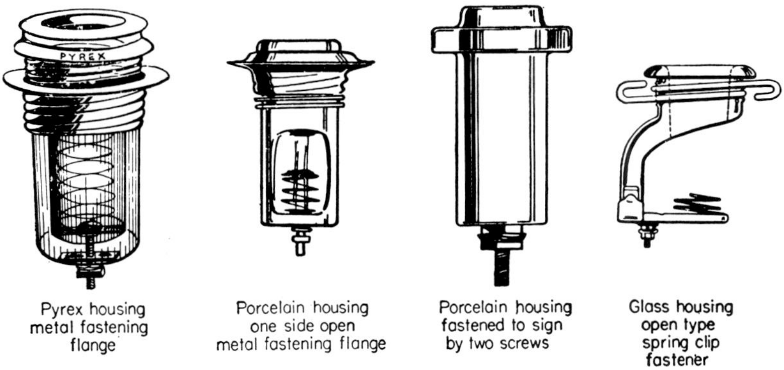

Figure 4.2. Electrode-insulator receptacles or housings.

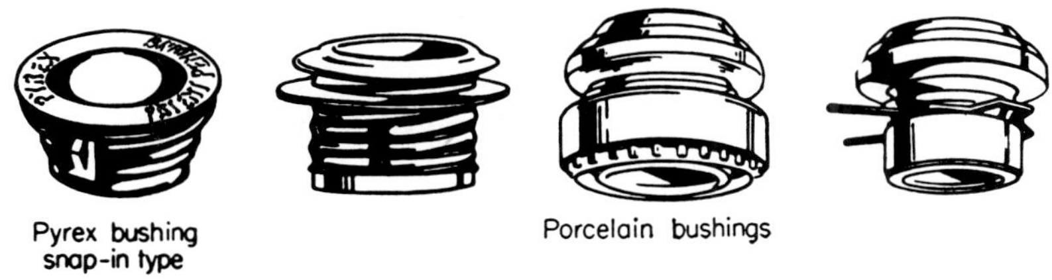

Figure 4.3. Electrode-insulator bushings.

extreme outer end of the housing is fitted with a brass binding post to which is attached the high-voltage cable. The Underwriters' Laboratories have set up standard specifications for these housings, and it is always advisable to use housings which have Underwriters' approval.

Electrode Bushings. In cases where a complete housing is not required, a bushing can be used which merely insulates the electrode from the metal sign. With this form of insulation, it is necessary to provide a flexible connection for connecting the electrode to the high-voltage cable. Cable-support insulators are used inside the sign box to separate the high-voltage cable and the metal box. Combined signs with bushings may be approved by Underwriters' Laboratories on the job.

High-voltage Cable. Attached to each electrode lead wire is a piece of high-voltage cable which connects it with the luminous

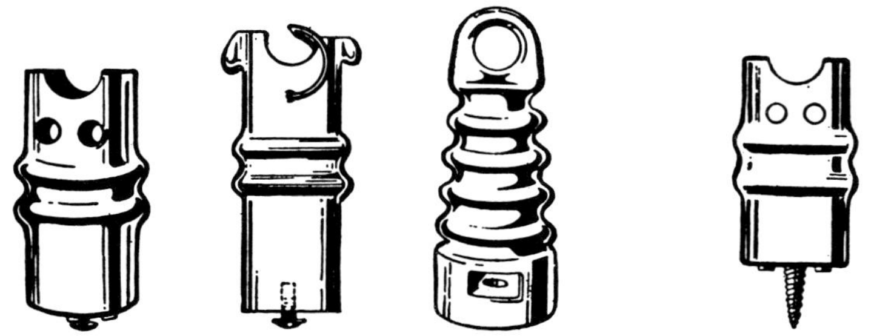

Figure 4.4. High-voltage-cable insulators or posts.

tube transformer. This cable must meet rigid specifications if it is to stand up under year-round weather conditions and the heavy electrical forces it must handle. Owing to the small current carried, the Underwriters have approved the use of one size copper wire, that is, No. 14, because of its tensile strength rather than because of its current-carrying capacity. The wire conductor which carries the current is made up of many strands of fine wire (usually 19), each covered with a thin coating of tin. The individual tinned wires are wrapped into one stranded wire to give needed flexibility to the cable. The wire can be coated with either rubber insulation or vinyl-polyethelene, a synthetic preparation. The insulation, consisting of a coat of polyethelene over the wire and an outer covering of vinyl compound, has properties far superior to those of rubber. At the same time, it will not dry out as rapidly under weather conditions. If rubber insulation is used, a number of layers of cotton braid may be woven over it. In some cases, fiberglas is used. The cable is then treated with a flame-retarding material.

Sign cable must withstand a very destructive effect known as *corona discharge*. When high voltage (15,000 volts) is applied to thin bare conducting wire in air, a flamelike glow of pale-violet color will surround the wire for 1/8 inch or so. This glow is called *corona*. It results from direct ionization of the air.

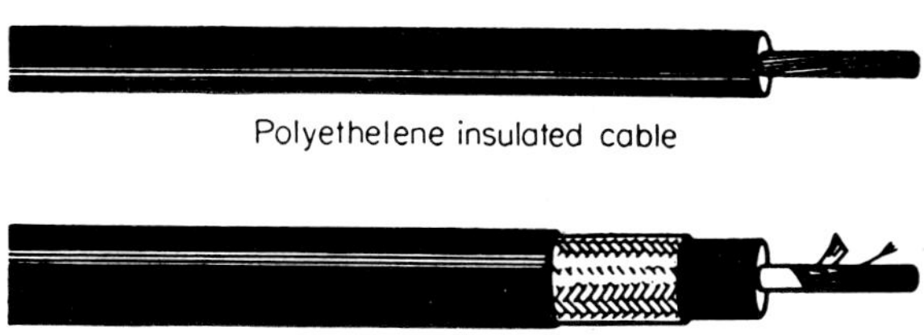

Figure 4.5. Two types of high-voltage cable used for neon signs.

Corona has a very destructive influence on poor insulating materials. For this reason, among others, even well-insulated ordinary electric wire is not suited to sign use. Only standard high voltage cable specially designed for luminous tubes should be used.

Since long lengths of high voltage cable increase the capacity across the transformer terminals, it may increase the terminal voltage of the transformer. This may cause harmful effects to the wire or to the insulation of the transformer. The smallest amount of cable required to do the job conveniently should be used. Long cable lengths must be avoided. Long cable lengths must be avoided particularly when running through metal conduits where the capacity effect is greatly magnified. Transformers should be brought as close as possible to the ends of the tubes.

Cable is available for use at three different maximum voltages. The 5,000-volt cable is identified by a red thread in the wire strands; 10,000-volt cable, by a yellow thread; and 15,000-volt cable, by a blue thread. The voltage rating of the cable should never be exceeded if satisfactory service is to be expected.

Many other forms of high voltage cable are available including lead-shielded and other types for exposed locations.

Electrical Transformers. The transformers used in electrical practice consist of a *core,* or central portion, made up of thin sheets of silicon steel. This core usually is in the form of a rectangle, with a rectangular cross section. A typical cutaway section of transformer core and coils is shown in Figure 4.6. On the core are wound two or more windings of insulated wire, each winding consisting of many turns. The electrical action of any transformer consists in stepping up or stepping down the voltage. That is, if a voltage is applied to one winding, another voltage will appear on the other winding, although they are insulated from each other. If the second winding has more turns than the first, the voltage across the second will be greater than that across the first; if it contains fewer turns, the voltage will be less. By proportioning the number of turns properly, the step-up or step-down ratio of the transformer can be made to suit any need.

The action of the transformer depends upon the magnetism set up in the core by the primary winding. The lines of magnetic flux in the core link the primary and secondary together, and as a result a voltage appears in the secondary, provided, of course, that the voltage is alternating. *A transformer will not operate on direct current. If connected to direct current, it will burn out almost immediately.*

Tube Transformers. For luminous tube work a step-up transformer is required. The primary, or low voltage winding, consists of perhaps a hundred turns of moderately heavy wire. The secondary, or high-voltage winding, consists of many thousands of turns of smaller wire. For a 15,000-volt transformer, the secondary would have approximately 140 times as many turns as the primary.

Ordinary power transformers, not used in luminous tube work, are usually designed to possess what is known as good *regulation.* That is, the voltage across the secondary is maintained almost constant regardless of the current being drawn from the secondary. This is a desirable characteristic for ordinary power work. For example, if a distribution transformer supplying 115-volt power to homes did not possess good regulation, the lights in the house would become dimmer as more lights were turned on.

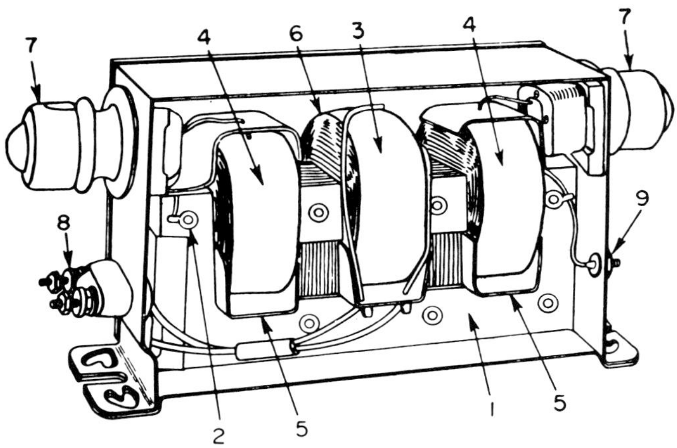

Figure 4.6. Cutaway section of neon transformer. 1. Low-loss laminations. 2. Lamination rivets. 3. Primary coil. 4. Secondary coils. 5. Insulation. 6. Compound for filling can. 7. High-voltage insulators. 8. Low-voltage connections. 9. Grounding terminal. (Acme Electric Corporation.)

In luminous tube work, however, good regulation is *not* desirable. This arises from the fact that the higher the current through the tube, the lower its resistance becomes. If a luminous tube were connected across a transformer having good regulation, the current would increase. As it increased, the resistance of the gas would decrease, and as a result, still more current would flow. In a short time, the current would become excessive and either the transformer would burn out or the tube would be destroyed.

What is needed is a transformer having *poor* regulation. That is, a luminous tube transformer should be built in such a way that the secondary voltage goes down as more current is drawn from the secondary. When connected to a gas-discharge tube, such a transformer soon reaches a condition of equilibrium; for if too great a current is drawn, the voltage goes down almost to zero, and the increased current can no longer flow.

Transformers of this type can be made easily if an extra magnetic flux path is provided in the core. This extra transformer leg is called a *magnetic shunt.* It acts as a safety valve on the transformer. As the current in the secondary increases, more and more of the magnetic lines of flux are bypassed by this extra shunt. As a result, fewer lines link the secondary winding. Thus, as the current in the secondary increases, the voltage across the secondary decreases and the required poor regulation characteristic is obtained.

The action of the transformer can be illustrated most clearly in graphic form. Figure 4.7 shows the voltage appearing across the secondary of the transformer for each value of current flowing through the secondary. When no current flows (the so-called open-circuit condition), the full voltage of 15,000 volts appears in the secondary winding. If more and more current is taken from the secondary, the voltage decreases. At about 25 milliamperes (the working current for which most sign transformers are designed) the voltage has dropped to a much lower value. Finally, if the two terminals of the secondary winding are connected together (the so-called *short-circuit condition),* maximum current will flow, which in this case is 30 milliamperes. Thus the secondary of a neon-sign type of transformer can be short-circuited without harm.

The dropping voltage characteristic of the transformer depends on the inductive action of the magnetic shunt. When the sign and its leads are connected to the transformer, the capacity

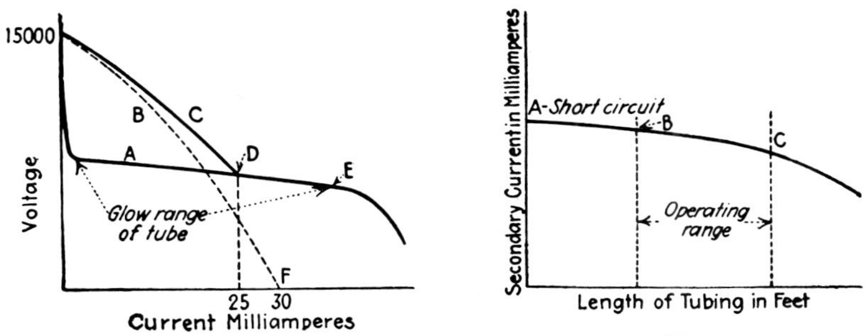

Figure 4.7. (a) Curves illustrating the action of a neon-tube transformer. A. Tube-characteristic curve, showing high initial breakdown voltage. B. Transformer-regulation curve (if load is noncapacitive). C. Actual transformer-regulation curve, shifted by capacitance of tubing. D. Stable operating point, due to current being limited by transformer regulation. Operating current at this point is approximately 17 per cent less than the short-circuit current shown at point F. F. Short-circuit current from high-voltage winding.

(b) The current-regulation curve of a typical neon transformer. Between points B and C is the safe operating range.

of these wires and the tube will neutralize some of the inductive reactance of the transformer. As a result, the voltage will drop off less rapidly as the current increases, as shown by curve *C* in the diagram (Figure 4.7). If lead-sheathed cable, conduit, BX, or other metallic covering is used over the cable, the capacity of the external line circuit will become still greater.[1] Under this condition, the voltage of the transformer may rise to higher values than are intended, and the transformer may burn out. It is recommended by transformer manufacturers that minimum lengths of grounded-conduit cable be used to prevent a capacity load of considerable magnitude which may increase the secondary voltage of the transformer.

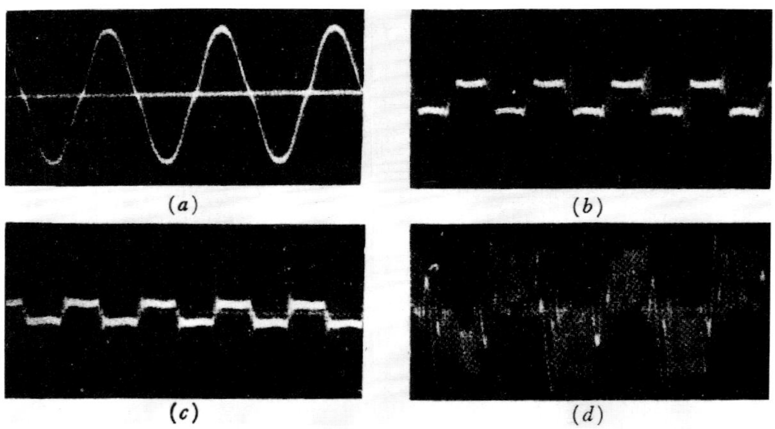

Figure 4.8. Wave forms of a 15,000-volt 30-milliampere transformer. (a) Secondary-voltage wave form with secondary winding open-circuited. The rms voltage is 15,000 volts and peak voltage is approximately 22,600 volts. (b) Wave form of correctly loaded transformer. The rms voltage is 9,000 volts and peak voltage is 19,000 volts. (c) Wave form of an underloaded transformer. (d) Wave form of an overloaded transformer. The peak voltage may reach more than 30,000 volts.

Increased voltage may also cause corona effects in the cable and breakdown of the cable or puncture of the high voltage insulation within the transformer itself, due not only to the higher secondary voltage, but also to the high frequency oscillations that are generated. This is illustrated in Figure 4.8, showing four different oscillographs.[1] Figure 4.8 (*a*) shows the secondary-voltage wave of a transformer with an open-circuited secondary.

[1]It has often been stated that it is injurious to operate a neon transformer for any length of time on open circuit, that is, without any load connected to its secondary terminals. It is true that the voltage under these conditions is higher than under operating conditions and that the insulation is thus subjected to a greater strain, but all transformers made by reputable manufacturers have a sufficiently large safety factor to take care of this condition.

Figure 4.8 (*b*) shows the same transformer operated with a normal length of luminous tubing. There is a pronounced distortion of the secondary voltage caused by such factors as the sudden ionization of the gas, the resistance characteristics of the gas, and the normal operating characteristics of the transformer itself. If shorter lengths of tubing are applied to the secondary of this transformer, that is, when it is underloaded, a wave form is obtained as illustrated in Figure 4.8(*c*). It is similar to Figure 4.8(*b*), except that there are fewer high-frequency oscillations and these are of lower amplitude. If, however, the length of tubing or load across the secondary is excessive, a wave form such as shown in Figure 4.8(*d*) is obtained. The oscillations are not only of higher amplitude than those in Figure 4.8(*b*) but are prolonged throughout each half cycle so they actually exist for a longer period of time. The peaks of the voltage represent rather high voltages which become even higher as a longer length of tubing is applied.

Comparing Figure 4.8(*d*) with Figure 4.8(*b*) shows that the combination of excessive peak voltages and high frequencies in the overload condition of Figure 4.8(*d*) causes a severe strain on the insulation of the transformer and the other parts of the high voltage circuit. For this reason signmen must take the utmost precaution to prevent overloading of sign transformers; otherwise transformer burnouts will occur, and frequent replacements will be necessary.

The magnetic-shunt or leakage-reactance type of transformer, as the type of transformer described above is called, has a very poor power factor. That is, the apparent power in volt-amperes as measured by the product of the primary voltage and primary current is much greater than the actual power in watts consumed. Since meters used in electric lighting circuits measure the true rather than the apparent power, this fact is of great importance only in special cases. In many municipalities, power factor of 90 per cent or better is demanded, and it is desirable to correct the power factor of neon transformers. If a standard low-power-factor transformer (approximately 50 per cent power factor) is in use at the time, it is necessary to correct the power factor to that prescribed by law. The method of doing this is described in a later section (page 79). Transformers can be obtained with power-factor correctors mounted in the same casing as the transformer. These are more expensive than the low-power-factor type but are less expensive than special capacitors designed for use with low-power-factor transformers.

Transformer Ratings. The power which a transformer can deliver is limited by its size. It is important to buy the size of transformer best suited to the job, as an oversize transformer is expensive in the first place and is inefficient when underloaded, while the undersize transformer will be overloaded and may cause flickering or strain and subsequent burnouts, as explained in the previous paragraph. Transformer ratings, which show the type of work for which the transformer is designed, must be considered carefully in ordering transformers from the manufacturer or supply house.

Transformers are rated by (1) primary voltage; (2) the power frequency, that is, 60, 50, or 25 cycles per second; (3) primary volt-amperes (the product of the primary volts times the primary amperes when operating into the rated load); (4) primary watts (the actual power consumed when operating at the proper load); (5) secondary volts (on open circuit); (6) secondary current in milliamperes for the correct operating load (called operating current); and (7) secondary current in milliamperes for short circuit (called short-circuit current). Transformers are often specified briefly by two values: their secondary voltage and short-circuit current, since these values will usually fix its size. All the above factors must be considered in designing the sign installation. The transformer-output ratings vary from 2,000 to 15,000 volts (secondary on open circuit) and from 18 to 120 milliamperes (secondary on short circuit). When the secondary voltage exceeds 7,500, the center tap of the secondary winding must be grounded.

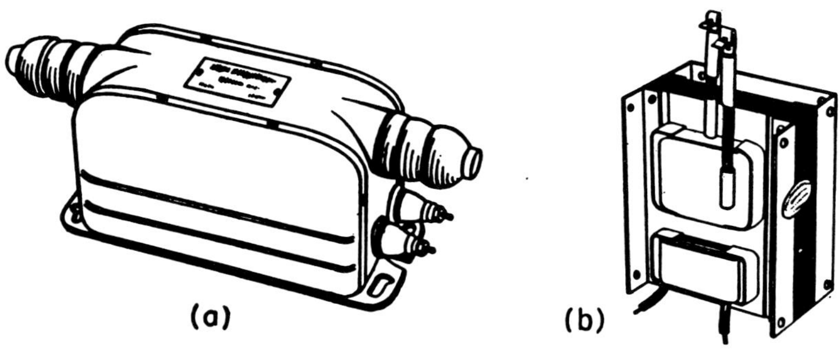

Figure 4.9. Two types of neon transformers in general use. (a) Standard outdoor-type transformer. (b) Core-and-coil transformer used mainly for indoor jobs.

For each size of transformer, there is a definite limit to the number of feet of tubing which can be handled. This limit depends upon the kind of gas used, the diameter of the tubing, and whether the tubing is filled with gas only or with a gas-mercury combination. The methods for matching the transformer to the tubing and testing for proper loading are treated in detail in Chapter 6.

It is always desirable to choose a transformer to fit the footage and to use nearly the maximum footage allowable for that transformer. In this way, it is fairly safe to assume that the transformer is neither overloaded nor underloaded. Actually, of course, the number of feet which can be operated from a given transformer is adjustable, since the correct current will flow almost regardless of the length of the tube, so long as the upper and lower limits are not exceeded. It is very important when designating the proper number of feet for a transformer to take into consideration the primary voltage existing at the job itself. Many signs may be properly designed according to the footage chart but may not operate efficiently or may flicker because the actual primary voltage at the source is *lower* than the rated primary voltage of the transformer. The secondary voltage is directly proportional to the actual voltage applied to the primary. If the primary voltage is low, the secondary voltage will be correspondingly lower. Therefore, a standard transformer carrying the rated number of feet of tubing may be overloaded because the primary voltage is lower than the rated value.

Physical Construction of Tube Transformers. The physical construction of the luminous tube transformer as shown in Figure 4.6 depends upon its intended use and also to a certain extent upon the manufacturer's design. The windings must be thoroughly insulated from each other. For this reason, layers of heavy paper, mica, or approved insulation are used between each layer of each winding. For low voltage types (2,000 to 5,000 volts), one magnetic shunt is usually used [see Figure 4.9(*b*)] and the secondary winding is completely insulated. For the higher voltage types (up to 15,000 volts), two shunts are used in a balanced construction and the middle point of the secondary winding is connected, or "grounded," to the case of the transformer. In this way, the maximum voltage of either high voltage terminal is never higher than half the terminal voltage aboveground. This reduces the strain on the insulation and balances the high voltage circuit.

The coils wound on the core are fitted with leads and the whole transformer is inserted in a metal case. The case is then filled with an insulating compound which completely encloses the transformer. This compound provides insulation, keeps moisture away from the windings, and assists in keeping the transformer cool. The metal case is fitted with low voltage terminals which connect to the low voltage primary winding and with high voltage bushings to which the high voltage winding is attached.

The exact form of the transformer depends upon the type of service for which it is intended. The standard enclosed type is mounted inside the metal sign housing. A suspension type is supplied with high voltage cable, low voltage wires, and a pull-chain switch for use in a store window or other display. The electrode-receptacle type has special housings, not unlike the porcelain housings already described, into which the ends of the tube may be fitted. In this type, the transformer case takes the place of the metal sign box and is often colored to improve its appearance.

The *core-and-coil* type of transformer is not mounted in a case and for this reason is considerably cheaper. Such transformers are used only in indoor installations where they can be thoroughly protected both from the public and from the weather.

They are provided with voltage ratings of 2,000 to 7,500 volts but not higher. This type of transformer has secured the component listing of the Underwriters' Laboratories. It should be used in a completed sign unit and protected in such a way that it cannot be tampered with.

Power-factor Correction. As stated previously, laws have been passed in many municipalities which require high voltage transformer installations to maintain a power factor of 90 per cent or better. These laws have focused the attention of the luminous tube trade on the question of power-factor correction.

In direct-current circuits, the power (in watts) supplied to an eletrical device is always equal to the product of the voltage (in volts) times the current (in amperes). In the alternating-current circuits used in luminous tube installations, the power is less than the product of the voltage times the amperage. To find the power supplied to an alternating-current circuit, the volt-amperes (the product of the volts and the amperes) are multiplied by the power factor. Thus, if a transformer draws 5 amperes at 115 volts, with 80 per cent power factor, the volt-amperes are equal to $5 \times 115 = 575$. The power is equal to 80 per cent of this, or $575 \times 0.80 = 460$ watts. The higher the power factor, the closer the wattage approaches the volt-amperes.

The desirability of a high power factor arises from the fact that the higher the power factor, the lower the line current, for a given input wattage to the transformer. Thus a 500-watt installation on a 115-volt circuit draws 4.8 amperes from the line at 90 per cent power factor. At 50 per cent power factor, typical of many uncorrected transformers, the line current for the same installation is 8.7 amperes, or nearly twice as great. The increase of current causes a loss of power in the supply circuit, for which the customer does not pay since his meter registers only the watts consumed. The line losses must be borne by the power company, which can do so only by increasing the rate. Thus, in the interest of economizing power and maintaining a low-rate service, many power companies require that the power factor be kept above some specified minimum, usually 90 per cent. The legislatures in many states have backed the power companies in the interest of economizing electric power by passing laws prohibiting low-power-factor installations.

High voltage luminous tube transformers can be obtained in either the standard type or the power-factor-corrected type. The standard-type transformer has a power factor of approximately 50 to 55 per cent. This low value is due to the magnetic shunt, which limits the output current by shifting the magnetic lines of flux as the load changes. This shunt introduces electrical inductance into the circuit, and this inductance causes the low power factor.

The power-factor-corrected transformer has been designed by the manufacturer to give a power factor of 90 per cent or better. This increase in power factor is accomplished by the

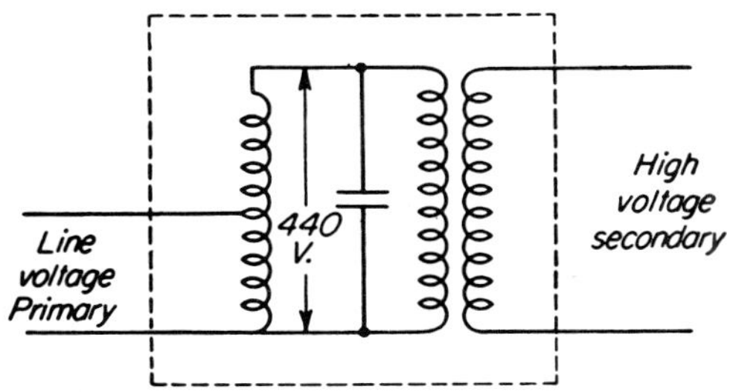

Figure 4.10. Schematic diagram of a high-power-factor transformer. The special primary winding and the capacitors are both placed in the same case with the neon transformer.

use of special windings and capacitors as indicated in Figure 4.10.

Power-factor correction on existing installations can be attained by the use of capacitors placed across the primary of the transformer.

To correct power factor by inserting capacitors across the primary as indicated in Figure 4.11, it is necessary to use a high value of capacitance. For example, to correct to 90 per cent the power factor of a standard 15,000-volt transformer producing a secondary current of 30 milliamperes and operating on a 115-volt primary circuit, there will be required a capacitance of approximately 70 microfarads. The same transformer designed to operate on a 220-volt primary, instead of 115 volts, will require only approximately 18 microfarads across the primary to give 90 per cent power factor. A similar transformer operating on a 440-volt primary circuit will require only about 4 microfarads. It can be seen from the example above that as the primary voltage is increased, the amount of capacitance necessary to correct to the proper power factor is materially reduced.

Capacitors for power-factor correction are rated in microfarads, and range from about 1 to 50 microfarads. The higher the power of the installation, and the lower the power factor, the larger the value of capacitance required to correct the power

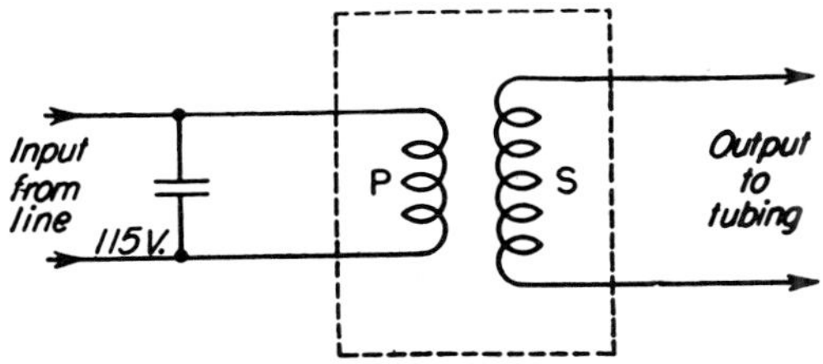

Figure 4.11. Power factor of a transformer can be corrected by placing a condenser across the primary line.

factor. The capacitors used should always be of the alternating-current type. The voltage rating of the capacitor must be rated for the primary voltage for which it is intended. Correction of power factor on existing transformer installations is always applied on the primary side of the transformer, and the capacitor should be connected directly to the primary terminals of the transformer, not to a fuse box or other outlet feeding several circuits.

Rotary Converters for Direct-current Operation. Alternating current, required for every high voltage luminous tube sign, is almost universally available. In rare cases only direct current is provided. In such cases, it becomes necessary to employ some sort of machine for converting the direct current into alternating current. The most common machine for doing this is the *rotary converter,* so called because it is a rotating machine (closely resembling an electric motor in outward appearance) which converts direct into alternating current. The rotary converter has two sets of terminals; the input terminals are connected to the direct-current supply, while the output terminals, which provide the alternating current, are connected to the tube transformer. The converter rotates as a motor, and its speed of operation depends upon the direct-current voltage which is supplied. If this voltage is lower than rated, the speed of the converter will be below normal. If this is the case, the frequency of the alternating current, which depends upon the speed of rotation, will be lower than normal. If a transformer is operated on alternating current of lower than its rated frequency, the transformer will take excessive current and may burn out; in any event, the operation will be uneconomical. Hence it is important to see that the direct-current supply is up to normal and also to see that the converter is not overloaded. Measurement of the input direct-current voltage with a voltmeter and of the speed with a tachometer is recommended.

The converter must, of course, be large enough to supply the load that it must carry. Converters are usually rated in volt-amperes, the product of the output alternating-current volts times the output alternating-current amperes. The sum of the volt-amperes of all the transformers to be operated from the converter must not exceed the rating of the converter; otherwise it will be overloaded and will not come up to proper speed. Most converters are built so that they require practically no attention except occasional oiling of the bearings.

If the transformer which loads a rotary converter is of the low-power-factor type, serious arcing at the brushes and commutator of the converter will occur. To prevent this, two alternatives can be used. The first is the use of a special converter having separate direct-current and alternating-current windings (an expensive type of machine), or else the power factor of the transformer load can be corrected by the use of capacitors. The latter method is usually to be preferred, because it is the most efficient. By the use of power-factor-corrected transformers which add very little to the cost, the volt-ampere capacity required of the converter can be reduced sometimes almost to one-half, and therefore the cost of the machine is considerably reduced.

Mobile Luminous Tubes. Another method of converting direct current to alternating current occasionally used for neon tube work is the inverter transformer. There are many of these in use and while they are no longer manufactured, it is believed that the reader should be acquainted with their operation. This transformer uses a vibrating-reed interrupter and is usually adopted for smaller transformers for operation from storage batteries. The average life of a reed unit exceeds 500 hours; the unit is easily replaced. There are two types: one for 6 volts and one for 12-volt battery operation. Inverter transformers will supply 6000 volts at 10 milliamperes.

Table 4-1. Inverter Transformer for 6- or 12-volt Battery Operation*

(Average safe operating footage)

	Red neon tubes						Mercury tubes					
Tube diameter, millimeters	15	12	11	10	9	7	15	12	11	10	9	7
Number of feet	20	15	12	11	10	9	25	20	17	15	12	10

* General Electric Company.

Table 4.1 illustrates the safe average operating footage for this type of unit. Inasmuch as alternators have replaced generators in most automotive applications, the production of AC 120-volt, 60-cycle is no longer a problem for a mobile application. Both inverters which transform DC to AC are available in larger capacities as are DC 6 or 12-volt motors which can turn small AC generators and provide the proper AC voltage for operation. Applications of this principle have been applied to floats towed by bicycles operated by batteries powering neon displays. Care must be exerted to properly support the luminous tubes so as to prevent breakage, but with the proper use of an inverter or a motor-driven generator this type of novel display is quite feasible. Care should be exerted to be sure that the frequency supplied by the generator or inverter is reasonably near 60 cycles or the nameplate frequency on the transformers used. This is also true of the voltage. It should also be remembered that high-power-factor transformers will appreciably cut the size and cost of the generating or converting equipment.

CHAPTER V

The Types Of Signs

Sign Backgrounds. Besides the luminous tube and the electrical equipment necessary to light it, there is another fundamental part of the luminous tube sign which has received only passing attention thus far. It is the sign background, which shows off the tubing to best advantage. A good part of the sign-making business consists of designing, building, and erecting these sign backgrounds. The development of interesting design layouts for signs is one of the essential factors in this industry. Although the many different types of sign backgrounds built today are similar to styles used for the past decade, the use of new ideas, the arrangement and kind of tubing, the color scheme, and the modernistic effect of the sign itself become important factors in the appearance of the present-day finished sign.

Electrical signs can be divided into two classes: spectacular signs and general signs. Spectacular signs are those which have been designed to attract immediate attention by some spectacular feature, such as those signs which are designed for Las Vegas' major casinos and the extravaganza in downtown areas, such as Times Square in New York City.

A good general sign background is designed by applying certain fundamental principles which make the sign not only attractive but also *readable* without undue glare. Excessive detail, small lettering, letters too close together, too many color combinations, and other factors which tend to blur the letters and fatigue the eye may cause an expensive sign to be a poor advertising medium. Therefore proper care must be applied in this phase which starts at the drafting table when the sign is first conceived.

OUTDOOR SIGNS

1. *The Swing Sign, or Projection Sign.* The swing sign, shown in Figure 5.2, is hung from a bracket so that it may swing freely. Usually a swing or projection sign is fitted with tubing on each face, in which case it is known as a double-face swing sign. Occasionally, only one face is fitted with tubing. The dimensions of the sign vary widely according to the particular installation. The metal sign box usually contains the transformer. In rare cases, the transformer may be mounted inside the building, but the long length of high voltage cable required makes this an undesirable type of installation. If the sign is of the flashing variety, both faces may flash together or alternately, one after the other. The

Figure 5.1. New York's Times Square is a focal point of vibrant neon activity at 42nd Street and Broadway.

Figure 5.2. Swing sign.

projecting sign is similar in construction to the swing sign, but the end is mounted flush on the side of the building without the use of outside guy bracing. A typical projection sign is shown in Figure 5.3.

2. *The Vertical Sign.* Figure 5.4 shows a typical upright or vertical sign, so named for its upright position. Such signs are usually longer and narrower than the swing sign and are mounted on several brackets attached to one edge. A combination of wire, chain, or angle-iron bracing is often used for support to the building. The transformers are mounted inside the metal cabinet. This type of sign is especially well adapted for large letters visible at a distance and for taking advantage of the vertical length of a building.

3. *Facia Signs.* Facia signs, sometimes known as belt or wall signs, are designed to lie flat against the building (Figure 5.5). They are usually signs of such height and length as to take advantage of the space available on the face of the building on which they are mounted. The construction of the metal box for a facia sign is similar to that for either the swing or the vertical sign with the transformers mounted inside the metal box.

4. *The Outline Skeleton Signs.* Long lengths of tubing without a metal box background are often used to outline the edges of buildings, canopies, faces of buildings or other permanent installations. The tubing is mounted either against a wall, background, or projection by use of long elevation posts. The electrodes may be protected by special housings, usually of the double

Figure 5.3. Projection sign.

type, so placed that there is complete drainage for rain or moisture. The transformers are placed either on the building itself in proper weatherproof boxes, or weatherproof transformers may be used. Short high voltage cable leads should be used to prevent high capacity effects. Sometimes transformers may be mounted on the inside of the building.

5. *Roof Structures.* Roof signs may be of many different types, depending upon the location and amount of space available. In general, a roof structure can be in either one of two classes. The open roof structure shown in Figure 5.6 consists of a framework of flat iron members properly designed from a structural point of view and thoroughly secured to the metalwork of the building itself. It is important to note that with a structure of this type, careful consideration should be given to the additional weight applied to the beams of the building. Reinforcement sometimes may be necessary.

A structure of this type is usually designed for a location where large letters are employed for long-distance display. The letters are individual units complete with tubing and including transformers. On very tall letters, the individual letters are subdivided into smaller units which are fitted together on the structure itself.

Figure 5.4. Vertical sign.

Figure 5.5. Store front showing facia sign.

Figure 5.6. Roof structure.

Another roof structure may be of the "billboard" type, which consists of flat board reinforced by a metal facing on which is mounted the tubing. The transformers may be mounted in metal cases or may be of the weatherproof type. In some cases reflectors are used. Roof structures may have a combination of luminous tubing, incandescent bulbs, and reflector lamps.

6. *The Pole Sign.* Any ground sign supported by one or more vertical pole supports is typically considered a pole sign (Figure 5.7). Such signs are often employed near curbside to identify business buildings that are set back from the roadway. A typical example would be a gasoline service station installation.

Figure 5.7. Pole sign.

The towering illuminated signs often viewed along today's expressway system identifying service facilities near the exits can also be considered pole installations, though in most cases, they are referred to as highrise signs. These displays are set on large diameter pipes or steel beams and may be erected to heights of 80 feet or more. While some of these displays employ neon for illumination, most are constructed with plastic faces and are internally lighted with fluorescent lamps. Some of the more recent installations may have porcelain faces and, in these units, illumination is provided by ground-mounted floodlighting.

Figure 5.8(a). The marquee sign, used by the theaters and restaurants.

There also is a third popular form of pole sign which is commonly designated as the pylon. In simple terms, this is a covered pole sign whereby the vertical supports are disguised behind decorative screening or sheet metal to enhance the aesthetic qualities of the total visual appearance of the installation. There is a decided concern for improved design quality in signage today and integrating the sign elements to harmonize with the colors, texture and architectural lines of the building identified. The pylon is an outgrowth of this trend.

7. *The Vestibule Sign.* The vestibule sign is mounted either outside the door of an establishment or inside the vestibule or in the place occupied by the door transom. The sign does not differ from the swing or facia signs except in its location.

8. *The Marquee.* The theater-type display shown in Figures 5.8(*a*) and 5.8(*b*) is known as a marquee. It is mounted on the special iron structure projecting over the sidewalk. It is almost always found in front of older theater entrances. In this type of

sign, the luminous tubes may be only part of the illuminated display. Incandescent lamps may be combined with the tubing. The removable opal-glass or plastic-letter signs such as are used to announce the picture currently showing at the theater may be surrounded by a decorative tube border. Fluorescent tubes are also being used in solid banks to illuminate the space underneath the marquee, the combination giving a brilliant display of color and light to illuminate the entrance to the building.

Figure 5.8(b). A typical marquee.

INDOOR SIGNS

1. *The Skeleton Sign.* The skeleton sign is a continuous tube formed into the desired reading matter. The glass tube itself is the sign without a solid metal background. In some cases, skeleton signs may have a skeleton metal frame for mounting the glasswork to reinforce the whole unit, preventing strains and providing additional strength. A skeleton sign may hang from either a metal box or a special skeleton type of transformer which is mounted on the ceiling. In the latter case, the sign should not be supported entirely by high voltage leads to which the electrodes are attached, but should have additional fine wires from either the ceiling or the metal box attached to the glasswork. In some cases, especially for small skeleton signs, the tubing is part of the hanging transformer and is directly attached to the transformer housing box so that the tubing and the transformer comprise one unit.

Transformers for skeleton signs are also available in which the two electrode housings form part of the metal container. A skeleton sign is made up and mounted to brackets attached to the case, while the electrodes fit into housings with spring contacts. As a rule, these units are not hung from the ceiling but are mounted on the window sill, counter, or any other available space.

Where a skeleton sign is used in combination with window-outline tubing, the sign may be suspended itself from the ceiling and suitably connected in series with the outline tubing from the same transformer. The tubing is so designed that the transformer will be connected to the nearest two electrodes.

2. *The Indoor Box Sign.* Although used only for directional signs, small complete metal boxes may be used for indoor mounting. The metal box is complete with tubing mounted on with elevation posts, electrodes insulated by small housings, and the transformer mounted inside the box. The whole unit is supported by cable, chain, or piping and may be either suspended from the ceiling or placed on the window sill, counter, or any other convenient indoor location.

3. *The Window Outline.* As explained previously under the skeleton sign, the window outline can be used by itself or in combination with a skeleton sign. It has become very popular because it attracts attention to the window while supplying illumination for it. The tubing is of outline type and is attached to the window by extension elevation posts.

4. *The Display Sign.* A popular method of using neon for indoor display is the form known as the display sign. These signs are small, usually a simple luminous tube border of one or more colors, perhaps with a letter or trademark worked into the tube. The border surrounds a message which can be inserted by the dealer and changed from day to day to meet the changing requirements of sales, menus, etc. Electric clocks are often mounted inside such neon borders, as shown in Figure 5.9.

Figure 5.9. An electric clock with ornamental neon tubing.

Figure 5.10. (a) The flat painted letter. (b) The raised letter. (c) Raised letter with shallow channel. (d) The channel letter. (e) Channel letters with multirow tubing. (f) The cutout letter.

Letter Types. The number of letter types that can be used for luminous tube signs are as numerous as the ideas of the designer himself. The luminous tubing will depend upon the type of letter that is going to be used in the sign. The signman must then take into consideration not only the height, but the width of the letter, so as to exploit fully the tubing used in the sign. The bending or treating of the box and the letters of the letter background to emphasize the luminous tubing is a very important factor in achieving the desired effect, not only when the tubing is lighted, but also when the sign is not lighted. The colors for the background should harmonize with the tubing and give an overall desired effect to attract the attention of the passerby.

The simplest type of letter background is the *flat letter* shown in Figure 5.10(*a*). This is simply a letter painted on the flat side of the metal box immediately behind the tubing. The metal box using flat letters is simple to construct, and a sign of this type is relatively inexpensive.

The *raised letter*, shown in Figure 5.10(*b*), is a simple metal letter raised above the surface of the metal box and having the same outline as the luminous tubing. This type of letter is easily

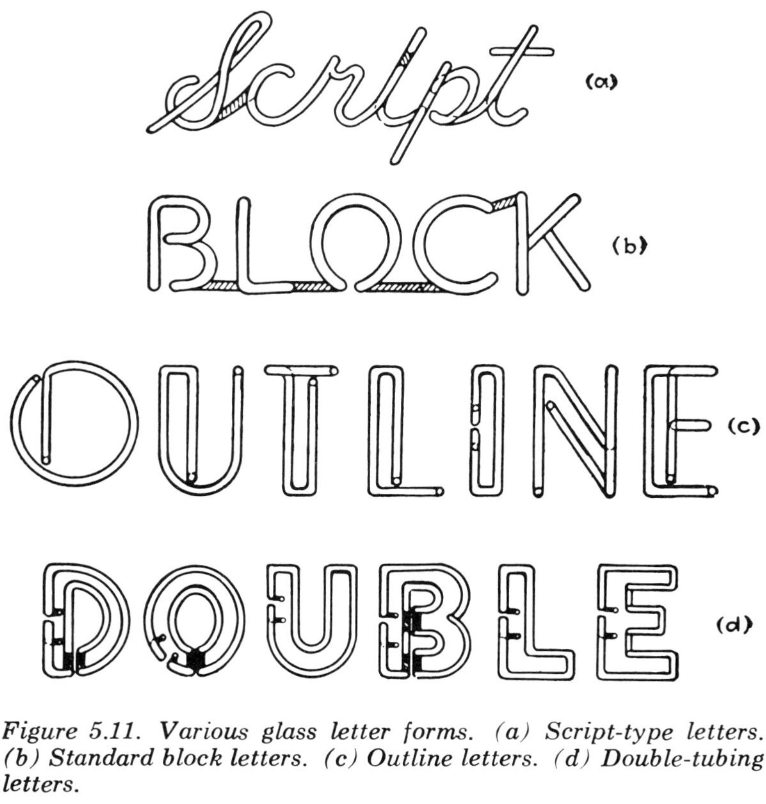

Figure 5.11. Various glass letter forms. (a) Script-type letters. (b) Standard block letters. (c) Outline letters. (d) Double-tubing letters.

seen in daylight and gives a finished appearance. In many cases, it is made of stainless steel or stainless steel finish. The raised letter may also be painted a different color from that of the metal box, and the same care in choosing colors should be used as in the case of the flat letter. The height, width, and general dimensions of the raised letters will depend upon the combination and design of the letter and tubing used, whether it is one, two, or more rows of tubing. Properly designed raised letters give a three-dimensional appearance to the sign, especially when viewed at an angle.

The *channel letter*, either Figure 5.10(*c*), Figure 5.10 (*d*), or Figure 5.10(*e*), is a raised letter made of metal. But instead of being flat on top, it has a channel in which the tubing is placed. Channel letter signs have a very clear-cut appearance and prevent the tubing of adjoining letters from giving a blurred and run-together appearance. The depth of the channel will depend upon the desired angle of visibility. If a sharp-angled visibility is desired, the channel must be relatively shallow. However, if a sign is to be viewed from straight ahead, with very little angle, then the channels can be much deeper to give a more striking individuality to each letter. Simple channel letters can be made either as shown in Figure 5.10(*d*), or they can be combined in a raised channel letter as shown in Figure 5.10(*c*).

The *cut-out* metal letter may often be used, as shown in Figure 5.10(*f*). These metal letters are usually placed on a box, approximately six inches away from the background with neon tubing placed behind the letters to outline them without the tubing itself being visible. The letters are then reflected from the background to get the desired appearance.

Figure 5.11 illustrates various glass letter forms used for signs.

Metalwork. The metalwork in making the sign cabinet involves bending the sheet metal into form, cutting out individual letters, drilling or punching holes, riveting, screwing, or soldering together the metalwork, etc. The art of metalwork is a science in itself, and many excellent reference books are available.[1] It is difficult in this book to go thoroughly into this subject, and therefore only a few details which may be of help to those making their own metal signs will be given.

The small, one-man sign shop generally uses hand tools to form the metalwork, drill holes, cut angle iron, etc. However, these are very laborious methods, and they take considerable time. As soon as a shop has a sufficient amount of work, it is important that the signman take advantage of the many different types of machines available for doing this work more quickly, more accurately, and less expensively. The metalworking machinery is not too expensive. Combination bending brakes are available for bending sheet metal up to 16-gauge and up to 121 inches long. The combination bending brake simplifies the accurate bending

[1]John C. Butler, "Sheet Metal Theory and Practice," John Wiley & Sons, Inc., New York, 1944.

of long lengths of metal. With use of box fingers and various attachments, all kinds of production work can be handled. There are deep-throat presses and hand punches available for punching all sizes of holes in metal. Both foot-operated and power-operated squaring shears are used for cutting long lengths of metal up to 12-gauge. Angle-iron tools for cutting, mitering, and bending angle iron will save considerable time by eliminating hand operations. A good metal shop should have hand seamers and notchers, electric drills, a bench grinder, an electric hammer, portable unishears for cutting out letters, and many other items which can accurately and efficiently meet the sign plant's needs.

Sheet metal varying in thickness from gauge 18 to 26 is universally used for sign backgrounds. The metals are of several varieties: galvanized sheet iron, "galvaneal" (patented process of treating sheet iron to prevent rust), stainless steel, enameled iron (sheet iron with a coating of enamel baked on it, similar to the finish used on automobile bodies), black iron, aluminum, and in some rare cases, burnished copper. Due to the scarcity of metals during the past few years, aluminum has been used to a greater extent than before and is almost exclusively employed in framing cabinets for plastic-faced signs. If the proper hardness of aluminum is used and the proper treatment applied, aluminum can be as good as any other metal in addition to having the advantage of lighter weight. Black Carrara glass ½ to 1-inch thick is also used as a background material.

In the construction of a metal box, it is very important to

Figure 5.12(a). Method of mounting transformer in sign box. Security of support, air ventilation, water drainage, and ease of removal must be considered.

provide angle or channel iron braces inside the box to give the necessary rigid construction. In fairly large signs, the sheet metal itself is not rigid enough to prevent bending or twisting. In smaller signs, channel braces are constructed of the same metal used for the face of the sign and placed inside the box to give the bracing required and provide proper support for fastening the sign to the face of the building. It also is important to provide special angle iron braces within the sign itself for properly supporting the transformer or transformers, at the same time providing ventilation and proper water drainage. Figure 5.12(a) and 5.12(b) show the method of placing and supporting the metalwork and the transformers.

Figure 5.12(b). This view is from the bottom of sign box and shows the transformer supports.

Sign-hanging Methods. Installing a sign so that it is mechanically secure is an art in itself. Many tubing manufacturers have found it wise to secure the services of professional sign erectors who are thoroughly familiar with the work, particularly if the sign is large and heavy. Sign hanging is a dangerous business unless carefully done, and in busy districts the hazard to the public is considerable unless the work is done by competent crews.

TYPES OF SUPPORTS

Swing Signs. The swing sign is supported at the top by brackets, which are built into the sign itself. From the forward edge of the hanging pole, guy stays are strung to the building wall. These stays run upward above the sign to the wall on either side. They may be made of wire, chain, or even of solid iron. The wire or chain stays are tightened in place with turnbuckles. The fastening at the wall is usually an eyebolt fastened with a lead expansion bolt. This bolt should be large enough and firmly embedded to support the sign safely, even when the sign is covered with ice or subjected to heavy winds.

Upright Signs. The bracing structure for upright signs is provided by channel or angle iron braces which fasten to the wall. The side bracing is also used to prevent swaying and should be very strong. Angle or channel iron cross bracing should be used within the sign itself extending out at a 45-degree angle to be fastened to the wall.

Facia Signs. The facia display is fastened flat against the building wall with expansion bolts. Stays or extra bracing are usually not required, but the fastening bolts should be strong and firmly embedded in the wall.

Roof Structures. If the area of the sign is large, the force of even a moderate wind will blow a roof structure over unless it is strongly built and well fastened to the roof. Installation methods will be determined by the construction of the building itself.

Projection Signs. The projection sign is a modified type of swing sign, projecting out toward the street with one side flush against a building. This type of sign is often built in combination with a facia sign, the two units blending together in a modernistic design.

The general procedure in sign hanging is as follows: A sketch is made of the sign and its supports, and the weight to be carried by each support is estimated or computed. For safety's sake, the actual weight of the sign should be multiplied two or three times. If the sign hangs down directly from two vertical supports, each support carries half the weight. If more supports are used, the weight is divided among them. This is true only if each support actually carries its full share of the load. If chain or cable supports are combined with solid angle iron supports, it often happens that most of the strain is taken by one support. If the chain or cable shows any slack, it is not carrying the load and will serve only as a safety measure in case the other supports fail. Chain or cable supports are suitable, however, if so placed that the weight of the sign fully takes up any slack. Cantilevered construction, however, is now favored to enhance the aesthetic values of the sign. Many newer sign ordinances prohibit chain and guy-wire elements because of their unsightliness and alternate construction and installation means become necessary.

If the support member does not fall vertically but is fastened back at an angle to the wall of the building, the strain on the cable increases. If the angle of the chain is 45 degrees to the vertical, for example, it carries about 40 per cent more load than if it hangs vertically. If the angle to the vertical is greater, the loading rapidly increases. For this reason the main support should not make a large angle to the vertical, preferably not more than 30 degrees. Chain supports at the side of the sign to prevent swinging carry much less load under ordinary circumstances. If the area of the sign is great, however, large side stresses will arise when the wind blows even at a moderate velocity. An occasional heavy blow may put a severe strain on the side guys; so in general they should be figured for as heavy a load as those which support the weight of the sign.

Chain and cable are available for supporting the weight of even the largest sign. The cable is ordinarily fastened to wall brackets or eyebolts, which are secured to the wall by lag screws.

These should be thoroughly anchored by the use of expansion bolts, or where possible by taking the bolt through the wall and securing it on the inside with a large washer and nut. A variety of shackles, wire rope clips (cable clamps), rope thimbles, and turnbuckles are available for finishing off and tightening cables.

The tensile strength of the sign hardware used in hanging the completed sign not only depends upon the item itself, but also upon the material into which it is secured. A lag screw may have a higher tensile strength than required for a job, but the expansion bolt into which it is fastened may not be properly secured into the brick or concrete, and therefore the whole constitutes a hazard.

In cities which require inspection of sign installations, it is desirable to obtain from the inspector in charge a set of requirements which must be met before the installation is approved. In any event, it is advisable to err on the side of caution in erecting any sign over a public thoroughfare, not only during the erection but thereafter during the life of the sign.

PART II

Shop Practice

CHAPTER VI

Designing The Sign

Sign-making Procedure. Of the many possible ways of going about the business of constructing a neon sign, the following general procedure has been found to be the most satisfactory. First, a preliminary or artist's sketch of the sign is made, more or less in detail but on a reduced scale. This sketch shows the message, the type of letters, the colors, and the style of mounting to be used in the sign. It is submitted to the customer for his approval. With it is often submitted a statement showing the cost of the complete sign, the manufacturer's guarantee, and an estimate of the cost of operating the sign.

If this preliminary sketch is accepted by the customer, a complete mechanical drawing or *working layout,* is drawn, which gives the details of every part of the sign, including metalwork, glass bending, electrical apparatus, and connections. This layout is drawn full size on heavy drawing paper. All dimensions, locations of drill holes, etc., are accurately located on it. The layout serves many purposes. From it are prepared the glass-bending layouts, the metalworking templates, and the tubing-connection diagram. From it is made a list of parts required, the number of feet of tubing, the type and number of transformers, the number of electrodes, etc. The layout is an almost indispensable aid in computing costs, both of materials and labor. And, finally, it serves as a permanent record of the sign and its construction, for use in repairing and replacement after the sign has been installed. For this last reason, the complete working layout should be carefully preserved at least until the sign is no longer in service.

After the working layout is completed, an asbestos glass-bending layout is made from it. The glass blower then bends the tubes to shape, using the asbestos layout as his guide. The metalwork is based on a drilling and mounting template, which is likewise prepared from the complete layout. The electrical apparatus is taken from stock, as indicated by the layout list. The completed tubes are pumped, bombarded and filled, sealed, aged if necessary, and mounted on the sign box. The electric wiring is then completed, and the sign, properly painted, is ready for installation. Each step in this procedure is taken up in the following pages.

The Preliminary Sketch. The preliminary sketch, or artist's sketch, is intended to illustrate to the customer the design of the sign as it will appear when completed. It may be very elaborate in colors, showing a detailed picture of just what the sign will look like. Such detailed sketches are very useful in capturing the customer's order by appealing to his fancy. The sketch may, on the other hand, be a very rough outline penciled in the salesman's notebook, or it may be merely an oral description of what the customer expects the sign to look like. In every case, the preliminary plan must show the message of the sign, the type of mounting, the colors to be used, and in most cases the style of lettering desired.

A highly effective way of presenting the appearance of the sign to the prospective customer is the use of sample letters made of clear and fluorescent tubing, fitted with electrodes, pumped and filled in the usual way, and connected by individual switches to a small transformer mounted inside a carrying case. These sample tubes are illuminated in the presence of the customer to obtain his reaction and to allow him to choose from among the many available colors. Preliminary sketches have also been made on paper with the use of fluorescent paints which are illuminated by a small portable ultraviolet or black-light lamp.

The Master Pattern or Working Layout. It is the duty of the layout man (in small plants this man may also be the owner, salesman, glassworker, and pump man) to take the preliminary sketch and to draw from it the complete, accurate, full-size layout (Figure 6.1) from which the metalworker, tube bender, and electrician may do their work. The amount of detail information which this layout contains varies widely in different plants, but it is recommended that it contain every possible fact about the sign. The layout man may confer with the other workers in the plant before deciding on the layout, but all the facts should appear on the layout sheet, regardless of who originated them. The layout may then be taken as the complete record for cost computation and subsequent repair work, without fear of omission of some vital feature.

With this idea in mind, it is recommended that the original layout contain the following items, either on the layout sheet itself or on specification sheets attached to it:

1. The design of the metal box, in full size, showing letter types; layout of metal trimming and ornaments; the metal supports for strengthening the box, such as angle, flat, or channel iron; and methods for fastening the entire sign to the building.

2. The location of all mounting holes for elevation posts, electrode receptacles, and cable supports.

3. The kind and color of either clear- or fluorescent-glass tubing for letters and borders.

4. The diameter of the tubing in millimeters.

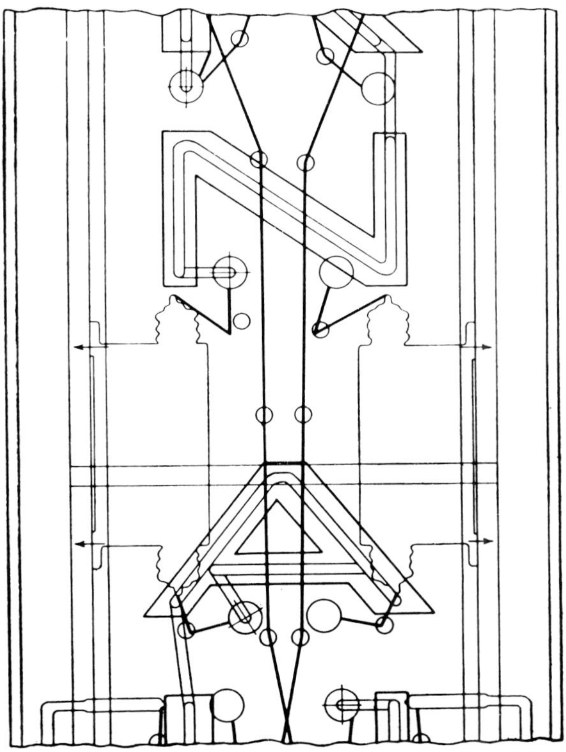

Figure 6.1. A section of a complete master pattern.

5. The length of each tubing unit, including bend backs and returns in feet and inches, showing the position of the electrodes.

6. The kind of rare gas required for each section of tubing and whether or not mercury is required.

7. The pressure of gas inside each section of tube.

8. The location and kind of electrodes to be used in each section of tube.

9. The type and location of each transformer, locating the necessary holes in the metal box through which transformers can be inserted or taken out from the inside.

10. The type and location of flasher, speller, or any other animating switch.

11. The complete wiring diagram, showing high voltage wiring, cable, jumper wires, etc., and the low voltage wiring, switches, time clocks, etc.

Designing the Tube. It is clear that the layout cannot be drawn up until the basic design of the sign has been developed from the artist's sketch. This depends upon the tubing. The tube design is the selection of the color and kind of glass to be used. When this has been decided upon, the first job is then to design the letters and borders, using the tubing selected which can be clear, colored, or fluorescent tubing. Table 3.10 can be used as a guide to select the kind of fluorescent glass necessary to give the desired color effect. To obtain the optimum value of gas pressure for neon tubing using clear glass and neon gas only and also for mercury and fluorescent tubing, both for signs and cold cathode lighting, Table 6.1 indicates the best values.

After the proper color glass has been chosen, the size of each section of tubing must be determined. Consideration must be given to the shape and length of the particular letter, not only for maximum efficiency but also for ease in handling when installing. Awkward sections of long lengths may result in excessive breakage. At the same time, the tubing units should not be made too short because each section and set of electrodes adds one foot to the total length of tubing. Too many short sections will decrease the amount of tubing that can be used on a particular transformer.

Then the proper diameter is chosen to fit a given transformer or number of transformers and Table 6.2 can be used to determine the transformer required for tubing length of a given diameter. A great deal of time and labor by the glass blower, in bending glass, can be avoided by the proper layout of the tube sections.

Please refer to Table 6.2, Transformer Footage Chart for 60-cycle Transformers. The chart indicates the normal maximum tubing operated based on average grade tubing. The first column on the left deals with the diameter of tubing to be used and indicates the gas pressure to be placed in the tube. The size of tubing has two lines running from left to right, one in red and one in black with white numbers. The white numbers indicate the number of feet of tubing which can normally be operated on the transformer of the voltage indicated in red on the line under the 25mm diameter tubing.

For example, down at the bottom of the second column there is a figure "102" in red. This indicates the normal amount of feet which may be operated on a 15,000-volt transformer of 25mm tubing filled with neon gas at the normal pressure. It also has a figure of 120 in black which indicates the number of feet of tubing which may be operated on a 15,000-volt transformer where the tubing is filled with a carrier gas and mercury as described elsewhere in the book. It would be noted that in the case of a 15,000-volt transformer the short-circuit might be 30 MA, 60 MA, or 18 MA, and the footage would remain the same. However, in the 12,000-volt column and in all the lower subsequent columns it will be noted that there are two sets of figures, at least as far down as the 15mm tubing between 7 and 15.

Table 6-1. Proper Gas Pressures for Sign and Cold-cathode Tubing

Neon-filled Tubes of Standard Length, at Least 4 Feet Long*

Diameter of tubing, millimeters	Transformer rating, milliamperes 30	60
	Neon gas pressure, millimeters	
7	16	18
8	15	17
9	14	16
10	13	15
12	11	13
13	10	12
15	8	10
18	7	9
20	7	9
25	7	8

Mercury-filled Tubes, Either Clear or Fluorescent, of Standard Lengths, at Least 4 Feet Long*

1. Use 6- to 7-millimeter No. 50 (B-10) gas pressure for all diameters for 30-milliampere transformers.
2. Use 7- to 8-millimeter No. 50 (B-10) gas pressure for all diameters for 60-milliampere transformers.

Fluorescent Tubes Using Neon Gas Only without Mercury, of Standard Lengths, at Least 4 Feet Long*

1. Use 5-millimeter neon gas pressure for diameters 12 millimeters and larger.
2. Use 7-millimeter neon gas pressure for tubes 11 millimeters and smaller.

Mercury-filled Cold-cathode Tubes Operating on High-voltage 100-milliampere Transformers or on 900-volt 120-milliampere Ballasts*

1. Use 5- to 6-millimeter No. 50 (B-10) gas or argon gas for diameters 15 millimeters and larger.
2. Use 6- to 7-millimeter No. 50 (B-10) gas or argon gas pressure for tubing smaller than 15 millimeters.

Cold-cathode Tubes of 25 Millimeters Diameter, 8 Feet Long, Operating on 750- or 800-volt 120-milliampere Ballasts

Use 4-millimeter argon gas.

* *Note:* If the tubing units are less than 4 feet long, a higher gas pressure must be used than that indicated above. We would suggest at least 2 millimeters more gas pressure above the values shown.

The first figure, for example, on 15mm tubing on a 12,000-volt transformer with neon is 45 feet. With an 18 MA transformer, and only 18 MA, the footage is 40 feet. Likewise in 7mm tubing on a 4,000-volt transformer for mercury-filled tubing, six feet of tubing may be used for a 60 MA or 30 MA, and the last figure always is the 18 MA transformer which has a slightly different wave form and is only five feet.

The important thing to remember is that one foot must be deducted from the footage chart for each pair of electrodes used. It is important to note that the resistance of any combination of tubing in any column is the same. In other words, each of the columns going down the page has the same resistance. What is meant by that is 21 feet of 7mm tubing is the same resistance as 102 feet of 25mm tubing filled with neon or 120 feet of 25mm tubing filled with mercury. This relationship is very important and must be understood in order to properly use the footage chart for loading transformers.

It will be noted that for each diameter of tubing and for each type of filling, be it with neon or be it with a mercury-filled tube, the resistance is different, and they can only be added when they are converted to the same resistance. Three examples are given below to clarify the use of this chart. However, before leaving the chart, it is desired to point out that 18 MA transformers are listed for most secondary voltages to the right of the 120, 30 and 60 MA listing. The 18 MA transformers are of a smaller milliamperage and will not carry the same footage. They are usually a core and coil or a window-type unit for quantity signs. They are not very common in higher voltages, but from 9,000 down they are quite frequently used for less expensive window signs where smaller diameter tubing is used and the smaller short circuit currents or operating currents are not a disadvantage.

Referring again to Table 6.2: Let us assume when we lay out a particular sign that the sign requires six units each of seven feet, and it is decided to make it out of 15mm glass. The decision for the diameter of the glass was made on the basis of the fact that it is going to be straight runs, there is no particular intricate bending, 15mm will be bright enough, and it will not be necessary to use more transformers required by smaller diameter tubing. Six units of seven feet each of 15mm red-filled tubing is a total of 42 feet of 15mm red neon-filled tubing. Since there are six units and according to the chart one foot must be deducted from the figures of the chart or added to the figures in calculating the footage to compensate for each pair of electrodes on each unit, this makes a total of 42 plus six, or 48 feet of 15mm red neon.

Referring to the chart, we find that on the 15mm line a 12,000-volt transformer may be used for 45 feet of 15mm red. However, we have 48 feet, and therefore we cannot use a 12,000-volt transformer. This leads us to the 15,000-volt column on the left which could handle as many as 60 feet of tubing and, therefore, is safe enough to use for 48 feet. Therefore, we would choose a 15,000-volt traansformer, and this 30 MA, 15,000-volt transformer is the correct answer. Higher current might be used for greater brilliancy, but this would not be the normal situation, as they are more costly to operate and more costly to purchase and, furthermore, would be unnecessary for anything except in very highly lit communities or ambiences.

Let us consider another example. Let us assume when another sign has to be layed out that it requires two units each of nine feet of tubing, 9mm in diameter, mercury and argon filled fluorescent green, for example, and also two units of eight feet each of 15mm blue tubing, mercury-argon filled. Now, two of the units of nine feet each is a total of 18 feet of 9mm tubing, plus two pairs of electrodes which is two more feet, 18 and two being 20. If you look at the chart on the 9mm line, there is a 20-foot figure above the 12,000-volt unit; and if this is dropped down to the 10mm line, the corresponding resistance is the same as 26 feet of 10mm glass. In the 11mm line, it is the equivalent of 30 feet; in the 12mm line, it is 34 feet; in the 13mm line, it is 40 feet; and in the 15mm line, it is 47 feet.

In other words, everything on that vertical line is the same resistance. Therefore, the resistance of 20 feet of 9mm mercury-filled tubing is the same as 47 feet of 15mm tubing. They are therefore interchangeable. Since we now have like things, we can add them. We are therefore considering the 20 feet of 9mm tubing equivalent for resistance purposes and calculation purposes only to be 47 feet of 15mm blue. In addition to that, there are two units of eight feet each which make a total of 16 feet, plus two pairs of electrodes, or 18 feet, making a total of 47 plus 18, or a total of 65 feet. Going along the table, we find that a 15,000-volt, 30 MA transformer will carry 72 feet. Since 72 feet is greater than the 65 feet we have calculated, we will use a 15,000-volt, 30 MA transformer for this assignment, as it will be well within the rating for the transformer. This method of transferring or calculating is used when two different sizes of diameters are utilized on the same sign.

Now, let us consider a third example. Let us assume we have a sign which will consist of 11 feet of 10mm fluorescent green, which is mercury-filled tubing with argon or No. 50, and also we have two units, one five-foot of 13mm red, and one four-foot of 13mm red. All these are to be illuminated by one transformer, and the following method is used to calculate the size of that transformer.

The five feet of 13mm red and the four feet of 13mm red, being the same diameter and the same gas filling, may be added together, making a total of nine feet, plus two feet for the two pairs of electrodes, since it is in two units. This is a total of 11 feet of 13mm red. Now, we must convert the 10mm fluorescent green into 13mm red in order to add likes together. This is done in the following manner. The 11 feet of 10mm fluorescent green

Table 6.2 TRANSFORMER FOOTAGE CHART

NORMAL MAXIMUM TUBING OPERATED . . . BASED ON AVERAGE GRADE OF TUBING

DIAMETER TUBING	GAS PRESSURE																			
7 MM	18 M.M.	NEON	21	16	15	12	10	9	8	7	5	5	5	5	4	4	3	2
		MERCURY	24	18	17	14	11	10	9	8	6	6	6	6	5	5	3	2
8 MM	17 M.M.	NEON	23	18	16	13	11	10	9	8	6	6	5	5	4	4	3	2
		MERCURY	27	21	18	15	13	11	10	9	7	7	6	6	5	5	4	3
9 MM	15 M.M.	NEON	27	21	18	15	13	11	10	9	7	7	6	6	5	5	4	3
		MERCURY	32	25	20	18	16	12	11	11	8	8	7	7	6	6	5	4
10 MM	13 M.M.	NEON	33	25	21	19	15	13	12	11	9	9	7	7	6	6	4	3
		MERCURY	40	30	26	23	18	16	13	12	11	11	8	8	7	7	5	4
11 MM	12 M.M.	NEON	39	30	26	22	18	16	14	13	10	10	8	8	7	7	5	4
		MERCURY	47	36	30	26	22	19	17	15	12	12	10	10	8	8	6	5
12 MM	11 M.M.	NEON	45	35	30	26	22	20	18	16	14	12	11	10	9	8	6	5
		MERCURY	54	42	34	31	26	24	21	18	16	14	13	12	11	10	7	6
13		NEON	50	39	33	29	24	22	20	17	15	15	13	11	10	9	7	5
		MERCURY	[illegible]	[illegible]	[illegible]	[illegible]	[illegible]	[illegible]	[illegible]	[illegible]	[illegible]	[illegible]	[illegible]	[illegible]	[illegible]	[illegible]	[illegible]	[illegible]

		15000	12000	9000	7500	6000	5000	4000	3000	2000
GAS PRESSURE 9 M.M.	MERCURY	72	55 47	40 34	28 25	23 20	19 17	16 13	12 10	7
18 M M GAS PRESSURE 8 M.M.	NEON	72	55	40	28	23	19	16	10	
	MERCURY	80	62	45	31	26	22	18	14	
20 M M GAS PRESSURE 7½ M.M.	NEON	78	61	48	34	28	23	19	12	
	MERCURY	90	70	55	39	32	27	22	16	
22 M M GAS PRESSURE 7 M.M.	NEON	85	67	57	41	34	28	23	14	
	MERCURY	100	79	67	48	40	33	27	18	
25 M M GAS PRESSURE 6 M.M.	NEON	102	79	67	51	40	33	27	17	
	MERCURY	120	95	80	61	48	40	32	22	

Voltage	15000			12000				9000			
SECONDARY Short Circuit MA.	60	30	18	*120	60	30	18	*120	60	30	18
SECONDARY Capacity VA.	900	450	270	775	720	360	225	600	540	270	180
PRIMARY Approximate Amps.	7.82	3.9	2.35	6.75	6.25	3.12	1.95	5.2	4.7	2.35	1.62
PRIMARY Approx. Watts Consumption at 115V. 60C.	450	225	135	715	350	180	110	550	270	135	90
Transformer Type	6015	3015	1815	12012	6012	3012	1812	1209	609	309	189

Voltage	7500				6000				5000			
SECONDARY Short Circuit MA.	*120	60	30	18	*120	60	30	18	*120	60	30	18
SECONDARY Capacity VA.	500	450	225	150	400	360	180	130	330	300	160	100
PRIMARY Approximate Amps.	4.35	3.9	1.95	1.3	3.5	3.12	1.5	1.13	2.88	2.6	1.39	.87
PRIMARY Approx. Watts Consumption at 115V. 60C.	450	210	110	75	290	170	80	65	300	140	80	45
Transformer Type	1207	607	307	187	1206	606	306	186	1205	605	305	185

Voltage	4000			3000			2000	
SECONDARY Short Circuit MA.	60	30	18	60	30	18	30	18
SECONDARY Capacity VA.	250	140	90	180	100	75	75	50
PRIMARY Approximate Amps.	2.16	1.25	.75	1.5	.87	.65	.65	.45
PRIMARY Approx. Watts Consumption at 115V. 60C.	110	60	40	85	40	35	35	23
Transformer Type	604	304	184	603	303	183	302	182

IMPORTANT: *120 M.A. ratings are P.F. corrected types. To calculate VA and primary amps. of power factor types not shown, multiply given ratings by .55. Deduct 1 foot from above figures for each pair of electrodes. Footages for mercury filled tubes are based on operation in temperatures above 40° F. Deduct 25% of footage for operation below 40° F.

plus one pair of electrodes, since it is one unit, is 12 feet of fluorescent green. The chart for argon-filled or No. 50 filled tubes with mercury for 12 feet of fluorescent green, 10mm, falls on the 6,000-volt, 30 MA transformer line. That is, the white figures with the black background. Remembering that everything on the 6,000-volt vertical line has the same resistance, going down to the 13mm red, we find that the equivalent resistance is 17 feet of neon-filled 13mm red; that is, only in resistance. Now, adding the 17 feet of 13mm red to the 11 feet of 13mm red, we previously calculated, gives us a total of 28 feet of 13mm red. Looking along the 13mm red column, we find that a 9,000-volt, 30 MA transformer will carry 29 feet. Since we only have 28 feet, we will use a 9,000-volt, 30 MA transformer for this problem.

The important thing to remember is that equal diameters, provided they have the same gas, can be added. Secondly, anything on any voltage vertical line may also be added whether it has the same gas or not, provided it is on the same vertical line. It is also essential to remember that every unit has two electrodes which must be added on to the length of the unit. This procedure permits the utilization of mixed tubing on a single transformer. Where the tubing exceeds the capacity of the transformer, more than one transformer should be used and a convenient break at the point of some number of units on one transformer established. The balance may then be placed on another properly chosen transformer.

The proper diameter is chosen to fit a given transformer or number of transformers and this diameter is balanced against the ease of bending small diameter tubing for small letters. For example, a three-inch letter would be practically impossible to bend out of 15mm or 18mm glass. By the same token, the other considerations are brilliancy against diameter which are treated in other sections of the book and the specific application of the tubing involved. For example, building outline tubing can easily be 15mm and in dark locations can even be 18mm where the brilliancy is not that important. On the other hand, as pointed out above, bending small skeleton letters or even small letters out of large diameter tubing to get more tubing on the transformer is impractical and must be ruled out. Each application must be weighed against the diameter of the tubing. Often an extra millimeter may be added to the tubing to fit the job conveniently on a transformer or set of transformers, but it is very doubtful that this possibility will be utilized for more than one or two diameters in size owing to the fact that the other considerations will more than outweigh the extra footage.

Metalwork Design. Once the tubing has been designed, the mechanical layout can then be prepared. For each electrode, the hole for an electrode housing or bushing must be located. Elevation posts should be indicated to support the tubing, at least two posts to a tubing section. It is important to place an elevation post as close as possible to each electrode to relieve the strain at these vital points. At the same time, all other holes for screws, mounting brackets, etc., can be indicated in the drawing. As soon as the size of transformer has been decided upon (see paragraph below), holes for mounting it should be marked in the proper place. In making the mechanical part of the layout, the thickness, or gauge, of the metal sheet, size of angle iron and metal supports should be given, and particular attention should be paid to the relation of these items to the mechanical strength of the entire metal box.

Electrical Design. The type and size of the transformer to be used depends upon the selection of the tubing, that is, whether it is filled with rare gas only or with rare gas and mercury, and on the length, diameter, and the current through the tubing. Using the layout sheet, or better still, the asbestos pattern, the total length of the tubing can be measured with a map-measuring wheel, as illustrated in Figure 6.2. If more than one diameter or

Figure 6.2. A map-measuring wheel for obtaining the length of tubing on a job.

kind of tubing is used on a job, each different section should be measured separately. When the total footage has been computed, Table 6.2 will indicate the type and size of transformer required. If the tubing exceeds the total that can be placed on one transformer, then proper division of the tubing should be made to use two or more transformers. Transformers should be placed inside the sign as close as possible to the two electrodes to which they are connected to avoid excessive length of high voltage cable. At the same time, the transformer should be mounted in such position as to provide ample ventilation and to avoid the accumula-

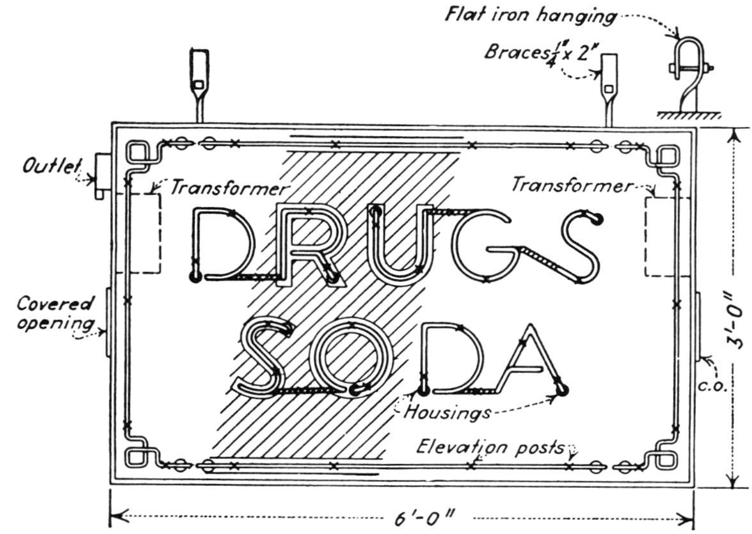

Figure 6.3. Complete sign layout, showing position of tubing, transformers, and details of metal drilling, etc.

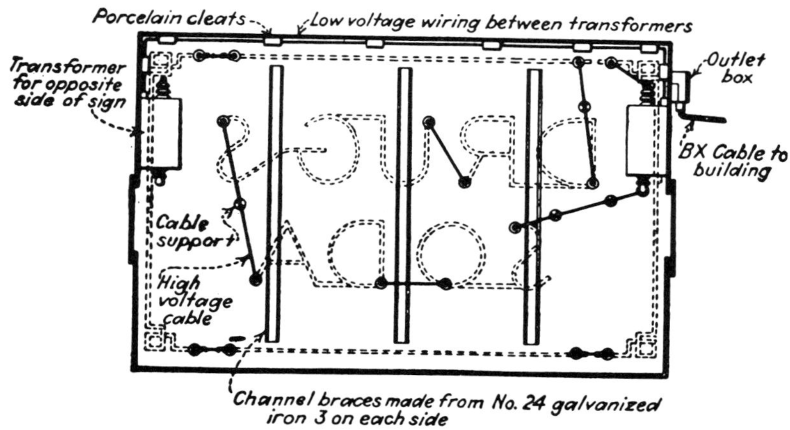

Figure 6.4. Wiring diagram of sign layout in Figure 6.3, showing one face of sign removed.

Table 6-3. Specifications for Sign Shown in Fig. 6-3

[Double-face swing sign reading: DRUGS (in red), SODA (in blue), with green border and flat-paint job]

Glass

Reading	Size of letter, inches	Glass tubing, milli-meters	Color	Type of elec-trode	No. of elec-trodes	Gas pressure, milli-meters	Mer-cury	Linear feet, 1 side	Linear feet, 2 sides	Cost of elec-trode	Cost of glass	Cost of gas	Cost of mercury	Total cost
Drugs	10	12, clear	Red	A15	8	12	No	12.5	25.0					
Soda	10	12, clear	Blue	A15	8	10	Yes	10.0	20.0					
Border		15, noviol	Green	A15	16	10	Yes	20.5	41.0					
							Total 1 side	43					Total cost	
								Total 2 sides	86					

Metal

Material	Type	Quantity	Cost
Sheet iron	No. 24 galvanized	62 square feet	
Outlet box	3-inch round	1	
Flatiron	¼ by 2 inches	8 feet	
Hanging pole	10 foot standard	1	
Guy chain	10 gauge 2/0	30 feet	
Expansion bolts	¼ by 2½ inches	3	
Sheet-metal screws	7 by ½ cadmium R.H.	78	
		Total	

Electrical Material

Material	Type	Quantity	Cost
Transformer	15,000 volts, 30 milliamperes	2	
Pyrex housings	TL-1	32	
Elevation posts	No. 1 glass	78	
Cable support	K	8	
High-voltage cable	15,000 volts	24 feet	
No. 14 BX cable	2 conductors	35 feet	
No. 14 weatherproof	Single conductor	20 feet	
Switch (metal box)	Double pole, single throw	1	
		Total	

Paint

	Quantity	Cost
Brush job:		

Table 6-3. Specifications for Sign Shown in Fig. 6-3 (Continued)

Labor Cost

Job	Hours	Cost
Glass bending		
Pumping		
Metalwork		
Electrical work on sign		
Electrical work on job		
Painting		
Assembling		
Hanging		
Sketching and layout		
	Total	

Total Costs

Material		
Labor		
Permit		
Maintenance		
Overhead		
Total		

tion of water. Also, the transformer should be mounted in a convenient location to facilitate removal for servicing on the job.

Where flashers or animators are to be used, the wiring diagram supplied by the manufacturer should be studied, so that the device can be inserted in the proper place to make the wiring short and convenient. When the location has been decided, the holes for mounting the flasher should be indicated in the proper place on the layout. When all the electrical devices have been located, the wiring diagram can be drawn into the layout chart.

With the complete facts and figures concerning the tubes, the metalwork, and the electrical devices on the paper layout, the layout may be considered complete, and the design of the sign is finished. The layout may then be used to draw up the specifications for the glass bender, tinsmith, and electrician, as outlined in the following chapters. A complete layout chart of a simple sign is shown in Figures 6.3 and 6.4. This diagram shows the correct layout for the complete design of the sign.

Preserving the Layouts. Some provision should be made in every shop for the preservation of each layout drawing after the sign made from it has been installed, since these layouts are the only accurate record of the signs that have left the shop. The layouts should be kept readily accessible at all times, regardless of when the sign was installed. They should be located in racks convenient to the use of employees yet out of reach of those who have no business near them. Each layout, template, and asbestos-glass layout should be tagged in some logical manner, so that it is not necessary to look through the entire lot to find the one needed. A company starting in the business should not forget to provide ample space for this sort of storage, because the drawings will accumulate rapidly.

Cost Computation from Layout Charts. The original layouts, metal templates, and glass-blowing layouts provide excellent bases for computing sign costs. Practically every item which goes into the construction of a sign can be taken from a well-drawn layout. From a list of such items, the cost of materials can be readily computed (see Table 6.3). Labor costs, both of tubing and of the metal box, can also be estimated from layouts, since the average time required to fabricate a given size box or a given footage of tubing can be found from experience. Even overhead costs can be estimated from figures taken from layouts, and in well-organized plants, the entire cost structure of the business is based on layouts and layout lists. For this reason, among the others given previously, every template and layout should be complete, and it should be kept on hand while any possibility of its being used again remains. It would also be of value to give the customer the cost of operating his sign. Table 6.4 gives the power consumed by various types of transformers. This should be useful in computing the cost of operating the entire sign.

The Factor of Experience in Layout Work. The layout man cannot do a good job on the complete layout chart of the sign unless he is familiar with the materials used in its construction.

Table 6-4. Power Ratings of Transformers* Uncorrected Power Factor

Secondary Volts	Secondary MA	Capacity Volt-Amp	Approx. Load Current (Amps)	Approx. Input Watts	Approx. Power Factor Per Cent
15,000	120	1650	13.75	900	55
15,000	60	900	7.50	495	55
15,000	30	450	3.75	225	50
12,000	120	1375	11.46	755	55
12,000	60	720	6.00	395	55
12,000	30	360	3.00	180	50
9,000	120	1080	9.00	595	55
9,000	60	540	4.50	295	55
9,000	30	280	2.33	140	50
9,000	30	280	2.33	140	50
7,500	120	900	7.50	495	55
7,500	60	450	3.75	225	50
7,500	30	225	1.88	113	50
7,500	30	235	1.96	117	50
6,000	30	180	1.50	90	50
6,000	30	200	1.66	100	50
5,000	60	300	2.50	150	50
5,000	30	150	1.25	75	50
5,000	30	160	1.33	80	50
4,000	30	120	1.00	53	45
4,000	30	140	1.17	63	45
3,000	30	90	.75	40	45
3,000	30	110	.91	49	45

Corrected Power Factor

Secondary Volts	Secondary MA	Capacity Volt-Amp	Approx. Load Current (Amps)	Approx. Input Watts	Approx. Power Factor Per Cent
15,000	120	900	7.50	810	90
15,000	60	450	3.75	405	90
15,000	30	250	2.08	225	90
12,000	120	775	6.46	700	90
12,000	60	400	3.33	360	90
12,000	30	200	1.66	180	90
9,000	120	600	5.00	540	90
9,000	60	300	2.50	270	90
9,000	30	150	1.25	135	90
7,500	120	500	4.17	450	90
7,500	60	250	2.08	225	90
7,500	30	125	1.04	113	90
6,000	60	200	1.66	180	90
6,000	30	100	.83	90	90
5,000	60	165	1.38	148	90
5,000	30	90	.75	81	90

*General Electric Co.

Unless he is an accomplished glass blower himself, he may indicate bend backs and special glass formations very difficult to carry out in practice. It is well, therefore, on complicated jobs, to have a conference between the layout man and the tube bender before the final glass layout is decided. The same principle holds true for the metalwork and the electrical wiring. Eventually, of course, the layout man will become familiar enough with the work to be able to decide upon the complete layout on his own authority, and he can then be responsible for the complete design of the sign.

The design principles given in this chapter are intended to guide only; the layout man may find that different procedures suit his particular plant much better than those given here. Nothing has been said about special design problems, such as obtaining special animated effects, high brilliance, and unusual color combinations. These advanced problems are treated in later chapters.

CHAPTER VII

Glass Bending

It is important to differentiate between glass bending and glass blowing. Glass bending is a comparatively new industry which grew up with the neon sign art. In glass bending, the operator heats and bends tubing into letters, borders, etc., according to a design on an asbestos layout. In glass blowing, the operator shapes and blows glass into various forms such as ornamental glassware, glass chemical apparatus, or any other articles requiring the technique of blowing and shaping glass. The difference between the two arts is that in one field tubing is bent and shaped to the same diameter and in the other case, glass is heated, blown out, and shaped. In both fields, the operator is called a glass blower.

The art of glass bending has also changed somewhat with the introduction of fluorescent-coated tubing. Previously, the glass bender of neon signs made a bend by heating the uncoated tubing in the fire, pushing and gathering the glass together, then pulling out and blowing the bend to the proper diameter. This was accepted as the method of making bends with the clear and standard colored glasses used at the time.

With the advent of fluorescent tubing, it has been found that excessive gathering of the glass at bends or heating the bends too hot caused a sintering of the fluorescent coating on the inside of the glass with the result that in some of the colors, such as blues and whites, the fluorescent properties of the powder were damaged and the bends appeared black. To avoid blackening at the bends, a revised method of heating the glass had to be evolved. The methods described in this chapter can be used for uncoated, clear, or colored glass.

Good glass blowing is almost wholly a matter of practice and experience; it is an art which cannot be learned by reading a book. The beginner cannot master the art of glass bending without continually practicing the different types of bends. It is simply a matter of practice, practice, and more practice. The more experienced one becomes in handling and manipulating heated glass, the better the workmanship and the faster the production.

General Shop Requirements. To do good work, it is essential that the glass blower have the proper equipment with which to work. Many good glass blowers are handicapped because of poor pressure regulation of the burners and resulting variation of the intensity of the flame.

The requirements for a well-equipped glass shop include the following: a supply of illuminating gas under constant low pressure, a supply of air under constant low pressure, a cross-fire burner or "cross fires," a ribbon burner, and a hand torch. For ordinary soft glass, these three types of fire—the cross fire, the ribbon fire, and the hand torch—constitute the required flames to bend glass into whatever shape is required. In addition to the fire equipment, a flat table is necessary, covered with a hard non-inflammable surface on which to place the asbestos patterns for bending the glass. A supply of heavy, strong asbestos paper for making layouts and a supply of glass tubing of various diameters are the essential materials required. Chalk or china-marking crayon for marking the glass and heavy carbon paper for tracing layout patterns should also be available. A supply of corks to fit the various diameters of tubing is also required (see Table 7.1),

Table 7-1. Cork Size Numbers Suitable for Plugging Tubes of Various Diameters

Cork Size Number	Fits Tube of Outside Diameter, Millimeters
000	8
00	9
0	10-11
1	12-13
2	14
3	15
4	16
5	17-18
6	19-20

together with various glassworking tools, such as a sharp file or glass-cutting knife, glass tongs, a rubber tube for mouth-blowing purposes, and a gauge for measuring tubing diameters.

Shop Layout. The layout of the tables in a shop depends almost entirely on the size of space available. Plenty of room should be allowed, particularly on the worktable, to take the largest piece of work to be handled by the shop. The straight glass tubing should be stored in racks near the table, since the closer it is to the table, the less time will be wasted walking to get it. The fires should be mounted on a separate bench about three feet from the long edge of the worktable and parallel to it. The burners should be mounted well apart so that a large piece of work will not get into two flames at once. A storage place for the asbestos layouts should be provided.

A convenient and compact layout for the fire table is shown in Figure 7.1. This table can be much longer or may carry more than one set of the fires indicated. On top of the table from right to left are a 16-inch adjustable ribbon burner, a five-way cross fire, and two torches, one a tipping torch for small work and the other a splicing torch for large work. The necessary piping between these fires and the mixers is shown below the top edge of the table. The fires are connected to their respective mixers by good flexible rubber hose varying in diameter from a small hose for the torches to a very large hose for the ribbon burner. On a ledge underneath the table is shown a small air blower connected to a motor. A relief-valve regulator is connected in parallel with

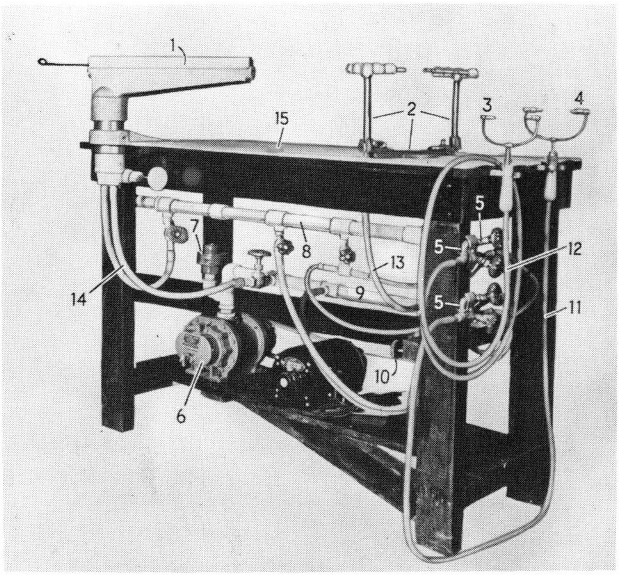

Figure 7.1. Compact layout of a fire table. 1. Ribbon burner. 2. Five-way cross fire. 3. Tipping torch. 4. Splicing torch. 5. Mixers for torches. 6. Blower and motor. 7. Relief value on blower. 8. Large-diameter gas piping. 9. Large-diameter air piping. 10. Mixer for cross fire. 11, 12, 13, and 14. Rubber tubing. 15. Asbestos paper or transite covering on table.

the piping, which provides for a constant rate of flow of air to the burners. Just above the air piping is the gas piping which may be connected directly to the gas main or may be connected to a gas booster. Although the air blower is shown directly underneath the table, it can be moved to a more convenient place if necessary. Should the noise from the blower annoy the operator, a muffler can be attached to the air intake of the blower to eliminate the nose. Both the gas and air piping should be of large diameter. A minimum diameter of 1½-inch piping is recommended.

The layout table, which is a flat bench close to the fire table, should be about 30 inches high and wide enough to take a 36-inch wide asbestos pattern. It should be level, to produce letters lying in a flat plane.

Buying Plant Equipment. When buying plant equipment, consideration must be given to the fact that nearly every shop finds it necessary to expand. The man entering the neon sign business should provide equipment which will allow for future additions. This is particularly important when purchasing air blowers or gas-boosting equipment. Extra fires can be added at comparatively low cost, but a new, higher rated blower or gas booster is a fairly expensive item.

Gas Supply for Burners. When installing a plant, it is advisable in the beginning to find out what gas is available for the burners. In many large cities and communities, manufactured or city gas is supplied to all points by gas mains running under the streets, and a connection can easily be made between the gas main and the shop. In other communities, natural gas may be available which is also piped, sometimes over long distances either from coal mines or oil wells. When neither natural nor city gas is available at a location, service is usually obtained by delivery of tank gas. In many Latin American countries where none of the above kinds of illuminating gas can be readily obtained, the manufacturer of neon signs may use a system using gasoline stored in a tank. An air blower forces a constant flow of air through the gasoline, creating a vapor which in turn is forced through the piping to the burners. Unless extreme precautions are taken to make such a system foolproof, this method may be dangerous owing to the presence of explosive gasoline vapors. Precautions can be taken by placing hand valves and automatic check valves in the piping system.

No matter what form of gas may be available, it must feed to the fires at a constant rate. Any variations in the rate of flow will cause the fires to flicker or change heating value. This condition requires constant fire adjusting and results in loss of time and poor workmanship.

City gas is supplied at a pressure which in most cases is satisfactory when used in a small shop. The pressure may vary from 8 to 14 inches of water column or $\frac{1}{4}$ to $\frac{1}{2}$ pound per square inch. However, in many locations the gas pressure may be considerably lower than the minimum required, or it may vary considerably because of changing gas requirements of neighboring consumers. Variation of temperature in the street may also change the gas pressure. As the weather gets colder, the pressure decreases in the gas mains. The same condition holds true for users of natural gas. The gas pressure may vary considerably with location and temperature. When fluctuation and loss of pressure are encountered, it is necessary to use a gas booster. Tank gas, which may be propane, butane, pentane, or any of the other mixtures available from oil products, is supplied in iron cylinders and compressed at a high pressure, usually 60 to 100 pounds per square inch. By means of a reducing valve, the pressure from the tank can be regulated and reduced to the ½-pound pressure necessary for operating the various burners. Figure 7.2 indicates a standard system employing tank gas.

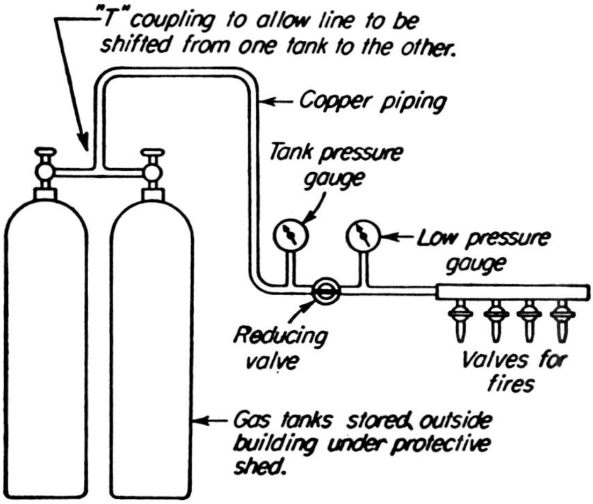

Figure 7.2. A tank gas system for providing fuel to fires.

Gas-pressure Maintenance and Adjustment. The pressure of the gas supply, especially natural gas, may vary from day to day, even from hour to hour. It is often wise to provide some method for maintaining a constant pressure. The most common way of maintaining gas pressure at a constant value is by use of a booster or gas pump. The booster consists of a mechanical blower, with tightly packed glands to prevent gas leakage, and a bypass valve regulator in parallel with the blower. The booster draws gas from the main line, increases and maintains a constant pressure to the burners, by bypassing through the valve regulator the excess quantity of gas.

Air Pressure. Air under a low pressure of one to two pounds per square inch must be mixed with the illuminating gas to produce in the burner the mixture required to give the proper flame. The extra heat is obtained by the mixture of the oxygen in the air with the illuminating gas. It is important that the air blower maintain constant pressure. There are a number of different types of air blowers available which operate on different principles. One type has a concentric rotor with sliding blades which by spring action fit tightly against the inside of the stator casing. Another type operates with two impellers shaped like figure of eights inside the casing. These impellers do not touch the wall of the stator but are closely machined to provide a space of 0.002 inch between the blade and the wall. The fast rotating action of the impellers provides a smooth flow of air from the blower. No oil is required inside this blower itself because there is no friction between the blades and the stator.

With any type of blower, an air-relief valve is usually furnished. This is illustrated in Figure 7.1. The valve is placed in parallel with the piping, and as the pressure tends to build up, more air will be bypassed through the relief valve to keep the pressure in the piping to the burners at a stable value.

In some blower installations, a tank may be required to give a stabilizing effect. Most of the modern air blowers do not require auxiliary air tanks because, with a suitable relief valve, a blower can be connected directly to the air line.

It has been the practice of many shops to use their high-pressure air compressors to supply air to the burners. This compressor usually is used for spraying paint on signs. The practice is to reduce the high pressure by means of a suitable valve. This arrangement is not satisfactory because there will be too great a variation in the flow of air since it depends upon the pressure existing in the tank.

To obtain satisfactory results from a standard low pressure air blower, a constant air pressure of one to two pounds is needed. To measure air pressure, a U-shaped tube filled with mercury is used. The length of the U tube can be about 12 inches. A difference of two inches between the two columns of mercury will indicate a pressure difference of one pound.

Glass-blowing Equipment. Cross Fire. The cross fire is most commonly used for making bends. When the glass blower becomes experienced, he may use the cross fire for making his bends, for sealing on electrodes, for sealing together his glass letters, and even for making tubulations. Some glass blowers seldom use the hand torches for a completed letter or section, making every operation in the cross fire.

It is for this reason that good cross fires should be purchased in the beginning. It has been found that cross fires with five tips on each side are most common and give satisfactory operation (Figure 7.3).

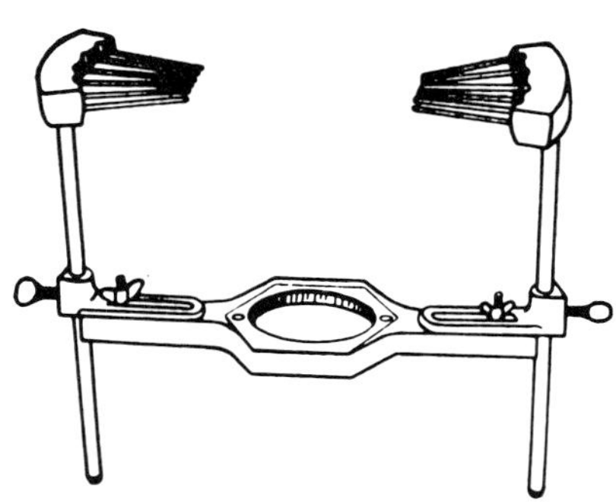

Figure 7.3. A standard cross-fire unit with five burners in each head.

Many experienced glass blowers who achieve high production sometimes use extra fires underneath the cross fire. This is done by use of either separate tips in a special unit or a short bunsen burner. Other glass blowers use either a six- or seven-way cross fire. Others get much better work by using a three-way cross fire. Another type of cross fire, shown in Figure 7.4, has one large circular tip on each side, spaced about two to three inches apart, opposite each other. This produces a hot flame of a diameter large enough to heat the glass for bending.

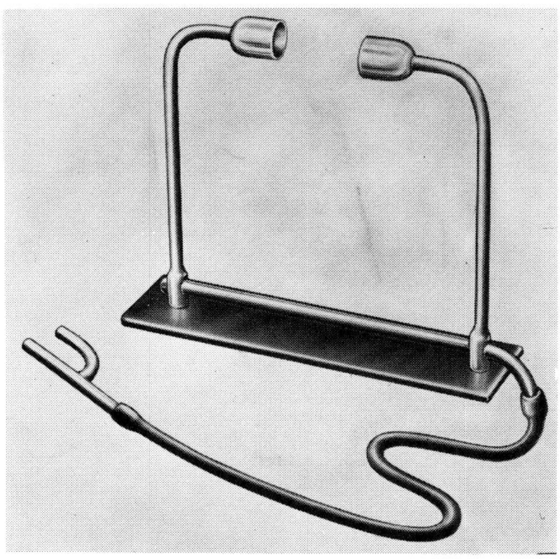

Figure 7.4. Cross fire with a single large burner on each side.

Ribbon Burner. The ribbon burner produces a solid ribbon of flame, as shown in Figure 7.5. It is used for heating a long length of glass and softening the entire length uniformly. The gradual bends of the curved letters O, C, G, B, R, etc., are made in this way. The flame looks like a solid layer of light blue flame

Figure 7.5. A ribbon burner for making circular bends.

just over the opening of the burner, with a large mass of dark blue flame extending for several inches above it. The width of the ribbon burner varies according to the kind of work.

Many different types of ribbon burners are available. The type shown in Figure 7.6 is practical and in general use. This is

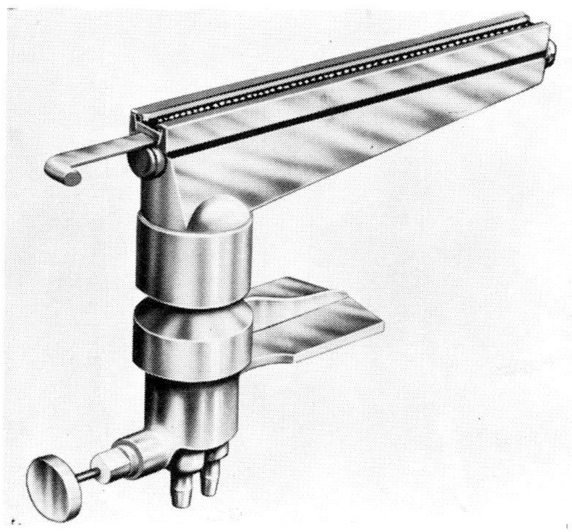

Figure 7.6. A standard 16-inch adjustable ribbon burner with built-in mixer. (Electrical Products, Inc.)

a ribbon fire giving a 16-inch flame of sharp and uniform intensity throughout its length. This ribbon fire consists of two parts: the mixer arrangement and the head which rotates on the mixing chamber. The burner head has an adjustable flat metal strip which can be moved in and out over the fire, allowing for adjustment of the length of flame from zero up to 16-inch full length. A six-inch burner head for small bends or a 24-inch head for longer bends can be interchanged with the 16-inch head, using the same mixer. There are other ribbon burners on the market, such as the one indicated in Figure 7.7. The tubular burners used many years ago have been superseded by the burners described above.

Figure 7.7. A fishtail type of ribbon burner.

Hand Torches. Hand torches have many different uses. The small hand torch, using burners having approximately a ½-inch-width flame, can be used for tubulation, for small splices, for use on the pumping system, and wherever a small, concentrated flame is required. The splicing torch has larger sized burner tips and produces a flame about ¾ to 1 inch in width and is used for splicing together pieces of tubing or for work where a wider flame is necessary. The standard type of hand torch is shown in Figure 7.8. Other types of hand torches are available which give a pointed, concentrated flame, as shown in Figure 7.9.

Sometimes a glass blower will improvise a small hand torch, as shown in Figure 7.9, by using a cross-fire tip. This is handy for making holes in glass tubing for attaching the tubulation. Another type of hand torch with a single flame is illustrated in the center drawing of Figure 7.9. This can be adapted for use not only with soft glass but also for hard glass of the pyrex glass type. This torch comes with the mixers attached directly to the torch itself and is also provided with removable tips for producing various flames.

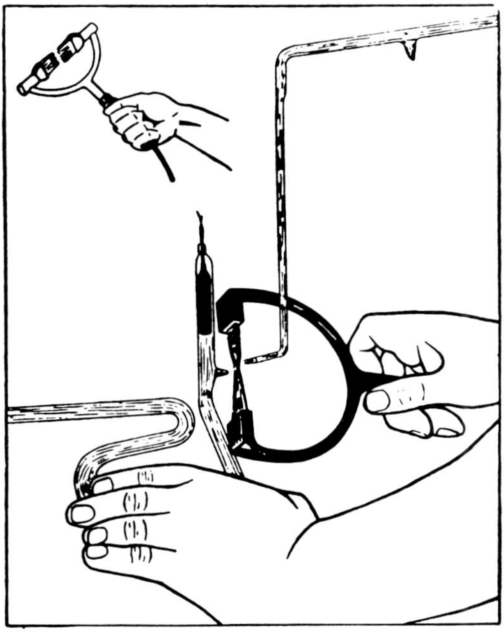

Figure 7.8. A small hand torch.

Cannon Fire. The cannon, or blast, fire, illustrated in Figure 7.10, is used by glass blowers who have had considerable experience in glass blowing. It provides a long hot flame which can be adjusted in diameter and length. The standard cannon fire has

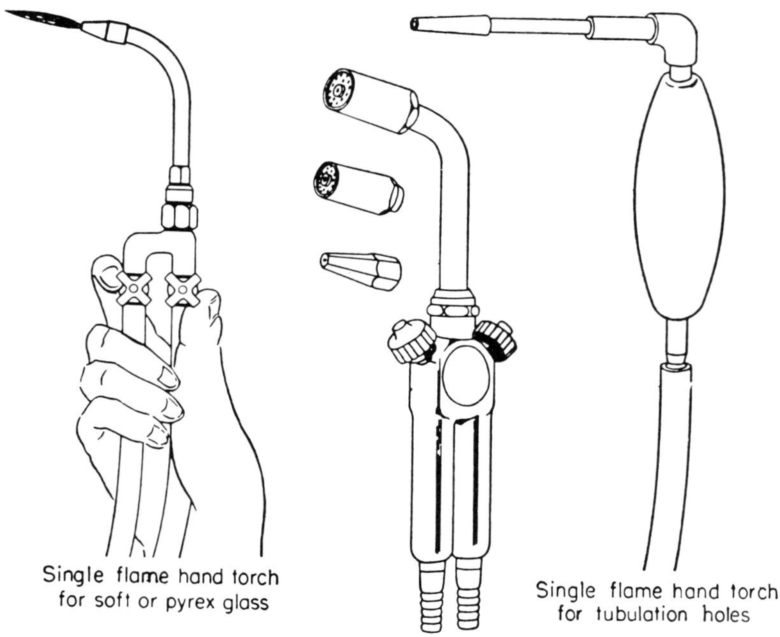

Figure 7.9. Three types of single-flame hand torches. The torch on the right uses a standard cross-fire burner.

Figure 7.10. A cannon or blast burner.

two inlets, one for air and one for gas. Other cannon fires can be obtained with three inlets, the third one for connecting to a tank of oxygen. By the addition of oxygen to the mixture of gas and air, a very hot flame is produced, necessary for melting and forming pyrex glass. There are glass blowers who use the cannon fire for making all their bends for the neon sign, especially when heating lime or soda glass.

Proper Burner Selection. A burner, whether it is a cross fire or any of the others described above, should never be ordered unless full information is given regarding the type of illuminating gas with which it will be used. Cross-fire tips, torch tips, and ribbon fires may look practically alike when compared with similar units, but there are slight differences in the diameters of the holes and the spacing used, which makes each burner more suitable for the type of gas for which it has been designed. Cross-fire tips designed for city gas will work very poorly when used with tank gas.

The Mixer. Every burner should have its proper mixer. The mixer may be attached directly to the unit as with some of the ribbons burners and hand torches, or it may come as a separate item. In either case, the mixer must be designed for a particular kind of illuminating gas.

Mixers usually have two valves, one for gas and one for air. In some cases, an extra valve is provided for oxygen. Whatever the gas and air, or the gas, air, and oxygen, in the proper propor- the design of the mixer, it is the function of the unit to combine tions to produce the necessary flame.

The quality or heating value of illuminating gas is measured in British thermal units (Btu) per cubic foot. City gas is the lightest of all and has approximately 540 Btu per cubic foot. Natural gas has a higher quality value, or approximately 1040 Btu per cubic foot. Propane (Pyrofax) gas has 2400 Btu, butane has 3200 Btu, and pentane has 4000 Btu.

Because the variation of the Btu value of the gas changes the heating value at the fires, the proper mixer must be supplied for a particular gas. For example, butane, or tank gas, has approximately six times more heating value than city gas. To compensate for this variation in quality of the gases, the mixer for each type of burner is designed to produce the proper proportion of air to gas. For city gas, it is designed to produce a mixture of approximately five parts air to one part gas. For natural gas, a mixture of 10 parts air to one part gas is required. For tank gas, 25 to 30 parts of air are required for one part gas. The mixing action is accomplished inside the mixer by use of a venturi injector which is designed to give the velocity required for each of the different types of gas (Table 7.2).

The Asbestos Layout. Before the work of bending the tubing can be started, the working layout must be made. This layout is drawn in reverse, that is, with each letter backward, as the sign would appear in a mirror. This is done so that the glass can be laid flat on the layout on which it is being worked and thereby

Table 7-2. Consumption of Various Kinds of Illuminating Gases When Used with the Fixed Amount of Air Shown, Using Different Types of Burners

Type of burner	Consumption of air, cubic feet per hour	Consumption of gas, cubic feet per hour				
		City gas	Natural gas	Propane (Pyrofax)	Butane	Pentane
Five-point cross fire	150	27.8	14.4	6.25	4.7	3.75
Small hand torch	40	7.4	3.85	1.67	1.25	1.00
Large hand torch	90	16.7	8.65	3.75	2.81	2.25
3N hand fire	40	7.4	3.85	1.67	1.25	1.00
12-inch fishtail ribbon burner (all out)	600	111.0	57.8	25.00	18.7	15.00
16-inch Hyde Kuert ribbon burner (all out)	1,115	206.0	107.0	46.3	34.8	27.9
Approx. ratio of air to gas	....	5:1	10:1	25:1	30:1	40:1
Btu values per cubic foot*	....	540	1040	2400	3200	4000

* Heat content from American Gas Association.

matched to the layout directly. The double backs, connections, and so forth will then appear at the back of the tube, rather than in the front, where they would be if a reversed layout were not used.

The layout is drawn on asbestos paper, since the hot glass may be pressed against such a material without fear of burning. Two methods are generally used in transferring the outlines of the letters from the original complete layout (see previous chapter) to the asbestos paper. The first uses a toothed wheel, which is rolled over the original layout, following the outline of the letters and producing a perforation or series of fine holes through the paper of the layout. The complete layout is then placed bottom side up on the top of the asbestos paper, and powdered charcoal is sprinkled over it. The charcoal leaks through the paper on to the asbestos and leaves the trace of the letters, in reverse. This trace can be strengthened with a charcoal pencil.

The other method of tracing the letter outlines onto the asbestos is the carbon-paper method, generally preferable to the perforation method. A piece of special black-carbon paper, as large as the layout, is placed carbon side up on the worktable. It is covered with the asbestos paper, and finally the complete layout drawing is placed face up on top of that. Using a heavy pencil, pointed metal, or glass stylus, the outline of the letters is traced on the original layout, and the carbon paper produces a reverse pattern on the asbestos paper. The pattern may be strengthened with charcoal pencil.

The glass blower's asbestos layout should be preserved as a record of the job after the sign is completed; it should never be used for another layout. Putting more than one layout on a sheet of asbestos is a common practice, but it is a false economy. Asbestos paper is cheap, compared with labor and other materials, and it should not be used too sparingly.

Matching the Layout. Before the glass tube is placed in the fires for making bends, it is marked off for length. Most sign tubing used comes in lengths between 46 and 48 inches. Some tubing is available in five-foot lengths, and tubing of larger diameters is available in eight-foot lengths.

The stick of glass is placed on the layout to see how many letters can be made from the length with sufficient space allowed for all the necessary bends. At the same time, at least six inches extra on each end should be provided to allow the operator to handle the stick.

When the length of tubing has been marked off in this manner, the bending begins. As each bend is made, the tube is matched to the layout while the bend is still soft, and the curves and angles are made to coincide exactly with the layout. In complicated letters like E, F, and H, the middle double back of each is made first, since this is by far the easiest way of doing the job. The bends on either side of the middle bend are then performed until the letter is complete.

For letters of 12 inches in height or larger, one or more 46-inch length of tube must be used for each letter, and splices must often be made in the middle of the letter. For smaller letters, often two or three letters may be made from one length of tube, the splices being made between letters where the painted-out portion appears.

The Influence of the Layout on Ease of Glass Blowing. It can be seen that the layout has a great deal to do with the ease with which the tube is bent into letters. If unnecessary double backs, awkward turns, or other errors of judgment are indicated in the layout, the glass blower's job is just that much more difficult. It is important, therefore, to use care and foresight in the layout of the pattern. The layout man may have a conference with the glass blower in order to decide just how the job can be done in the simplest and best manner. In general, the fewer the number of bends, provided that the letter outlines are still sharp, the better is the layout, and the longer a length of tubing can be made to go. Economy in the use of glass is a saving, but it should not be practiced if the glass blower's time is thereby increased, because the time is more expensive than the glass.

Bending the Glass. Every glass-bending operation requires the softening of a portion of the glass from a solid to a plastic state by heating in the fires. In this operation it is important to hold the glass by hand and rotate it properly; otherwise the distribution of heat will not be uniform and may cause twists, strains, or variation in wall thickness and result in an uneven bend.

When using a straight piece of glass, it is balanced so that all the fingers touch the glass tubing. The thumb is placed on top, and the four fingers support the glass from underneath, all touching the surface very lightly. The glass is rotated slowly by slight pressure from the thumbs above. This position is illustrated in Figure 7.11. The glass should not be held too firmly; otherwise too much pressure will be exerted on either side. The glass should be rotated slowly and evenly as it begins to soften, since in this way the operator will have better control and the fires will heat the glass uniformly.

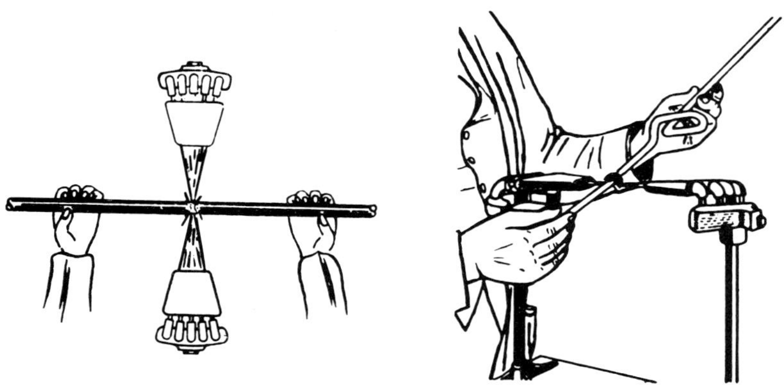

Figure 7.11. Holding the glass tubing in the cross fires.

When making bends in the cross fires, the amount of gas and air should be adjusted to allow for a slight fire splash on the glass surface which is at least ¼-inch wide. When electrodes are sealed to the glass in the cross fire, the fires can be adjusted for a sharp concentration of heat on the splice.

The present-day technique of bending fluorescent-coated glass is slightly different from that used years ago when only clear and solid colored glass were used. When a bend was made with clear glass, it was heated to the plastic stage, pushed together or gathered, and then pulled out slightly and blown to the proper shape. Such an extreme heat-treatment on coated fluorescent glass will cause the fluorescent coating to sinter, especially on blue and white tubing, and may affect the efficiency of the coating and possibly cause darkening. When a slight gathering is specified in one or two special cases described later, it should be done with caution and without too much heat being applied so as to prevent affecting the fluorescent powder.

In general, when working with fluorescent tubing, the bend should be made using as low an amount of heat as possible to bring the glass to a plastic state.

The Angle Bend. To make a simple right-angle bend of 90 degrees, as shown in Figure 7.12, the glass should be marked so

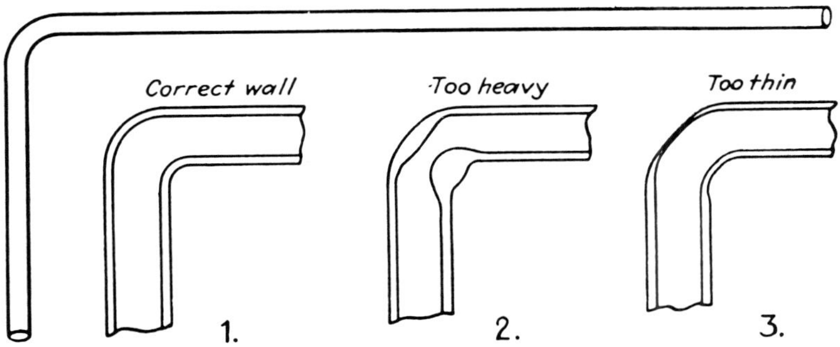

Figure 7.12. The angle bend. (1) A correct uniform wall bend. (2) and (3) Incorrect forms of bends.

that the portion of the surface which will be placed into the fire will be two full diameters of the glass. For example, if 12-millimeter glass is to be heated for an angle bend, the heating surface will be 24 millimeters in length and should be marked off by chalk. The glass is placed into the cross fire, rotated in the center of the fire, and, at the same time, moved evenly from side to side between the chalk marks until the glass becomes soft and plastic. Unless the operator balances the tubing, it will tend to give or bend. At this stage, the operator raises the glass slightly so that the fires hit the bottom of the glass or that portion which will be the outer surface of the bend. Then the operator stops rotating the glass and only moves it from side to side between the chalk marks. This will heat up the outer surface of the bend to a higher degree than the inner surface. When the glass has been properly heated, it is taken out of the fire and bent to the correct angle. By applying more heat to the outer surface of the bend, it is allowed to stretch to a greater degree, thus giving a more uniform bend. Also, if the inner surface is slightly cooler than the outer surface, the glass bender has better control and can make a more perfect bend.

Double-back Bend. The double-back bend is used for such letters as R, E, F, and G, where a complete 180-degree return bend is desired. This is illustrated in Figure 7.13. The portion of the surface of the glass which is to be heated for a double-back bend is approximately 2½ times the diameter of the glass. If the glass is 12 millimeters in diameter, the heating surface is approximately 30 millimeters in length and should be properly chalk marked. The glass is first heated to a plastic state by rotating and moving it from side to side in the center of the fire. It is then raised up so that the fire hits the bottom of the glass, and it is moved from side to side so that the outer surface of the glass is thoroughly heated while the inner surface is heated solely by conduction. When the glass is ready for bending, it is brought out of the fire, dropped about waist high, and the bend is made with both hands moving up at the same time. Care should be taken that there is no twisting or out-of-line motion when bringing up both ends of the glass. At this stage, when the bend is completed, it will be noticed that the outer surface of the double back is flat

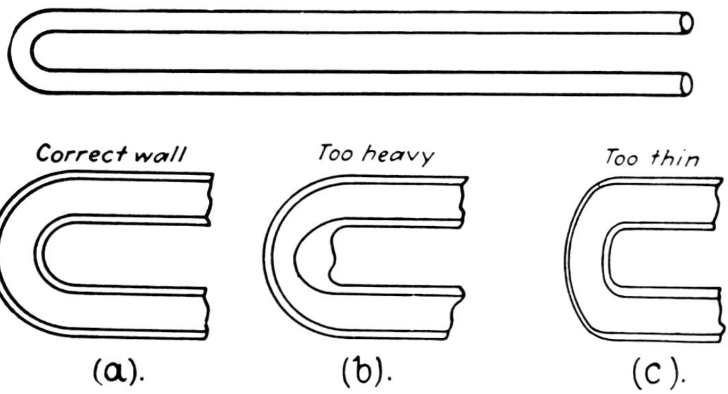

Figure 7.13. Double-back bends. (a) A bend with a uniform thickness of wall. (b) and (c) Incorrect bends.

while the inner surface is perfectly round. The outer surface is still in a plastic state, thus allowing the glass blower to blow out to the proper circular diameter. This is shown in Figure 7.14(*b*). When making double backs from glass of 15 millimeters or more in diameter, it may be well to deviate from the above procedure by gathering or thickening the glass slightly without applying too much heat. The glass should be gathered so that the outside diameter remains the same throughout the heating procedure, whereas the wall thickening is formed on the inside, as illustrated in Figure 7.15(*a*). The glass is then taken out of the fire, pulled out slightly [Figure 7.15(*b*)], bent into a double back, and blown out. Larger diameter glass, if not gathered slightly, will have a thinner wall than desired at the outside bend, as shown in Figure 7.13(*c*). Precaution should be taken to gather the glass only slightly, because if too heavy a thickness is formed, it will produce a crease, as shown in Figure 7.13(*b*), when finally shaped and blown out.

Combination Bend. A combination bend is one in which two bends may be made at the same time from one heating in the fire. Many different types of combination bends can be made, depending upon the layout of glasswork and upon the capability of the glass blower himself. However, there are a few fundamental combination bends which, if thoroughly known, will act as the basis from which all other complicated combination bends can be evolved.

The *straight-drop combination bend* is one in which two right-angle bends are made at the same time in a straight line, as shown in Figure 7.16. The glass is marked off for 2½ times the

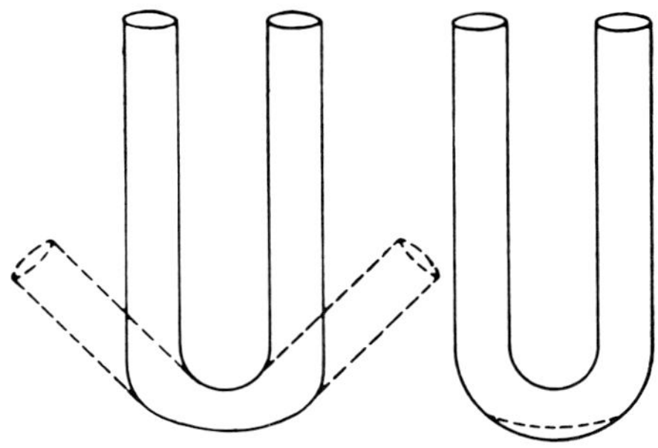

Figure 7.14. Double-back bend. (a) Showing flat at bottom of bend immediately after glass is bent. (b) After glass blower has blown bend to proper diameter and thickness.

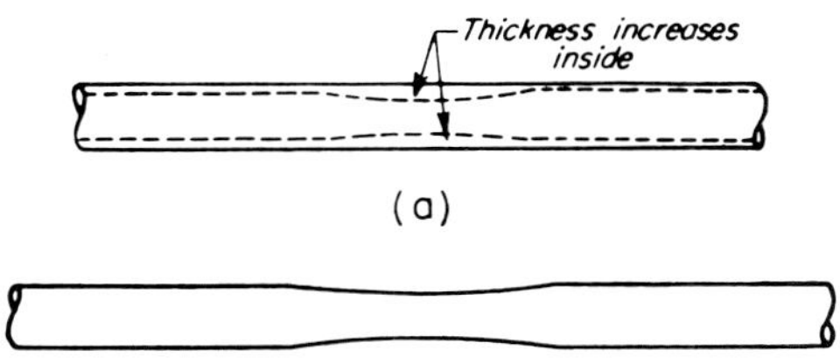

Figure 7.15. Formation of thickness of wall of large-diameter glass before making a double-back bend.

diameter for the heating surface. It is then rotated and moved in the fire until the entire surface of the glass between the two chalk marks has been thoroughly heated. As this requires two separate bends where part of the circumference of the glass is on the inside of one bend and the other part on the outside, it is necessary that the glass be kept rotating continuously and heated over the entire surface during the whole heating operation. This is a different procedure from either the angle or double-back bends. The glass blower should be careful to see that the entire surface is heated uniformly. When the glass is heated to the proper temperature, it is held straight out and one side is dropped to the height required for the bend and blown out.

Another form of combination bend is the *combination angle bend*, as indicated in Figure 7.17. This is used when making the center portion of the letter E or F. It consists of two bends, one which is a right-angle bend and the other a straight-drop bend, as indicated in Figure 7.16 by *A* and *B*. The glass is marked and heated exactly as in the combination straight-drop bend. When the glass blower is ready to make the bend, he first forms the angle bend *A*, as shown in Figure 7.18. He then brings down the glass to the proper drop and at the same time rolls that portion of the glass approximately 90 degrees inward toward the letter, as illustrated in lower diagram of Figure 7.18. The angle bend which was first made is still plastic enough to allow for

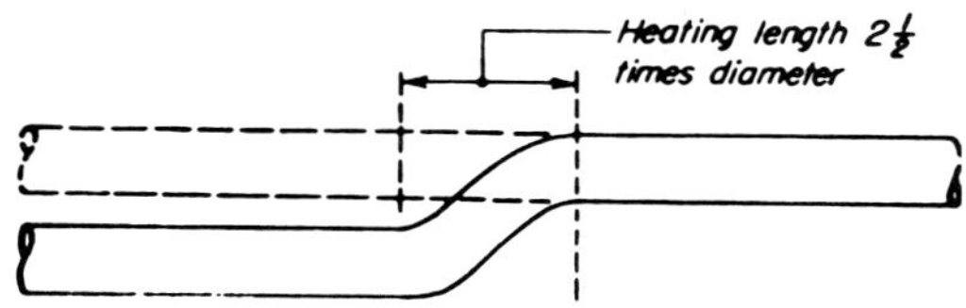

Figure 7.16. A straight-drop combination bend.

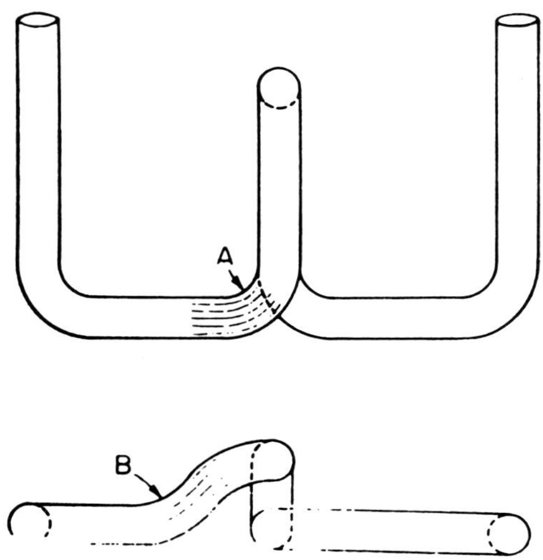

Figure 7.17. Combination angle bend. Both A and B bends are formed in one heating operation.

dropping down and rolling the glass. Both bends are made practically at the same time, although the explanation given above indicates two different operations.

The same technique as explained above can be used to make any type of combination angle drops. For making the letter T, as illustrated in Figure 7.19, the same procedure is employed, except that a longer surface of the glass must be heated up for a proper combination bend without stretching the glass too much.

Ribbon-fire Bends. A ribbon-fire bend, in which a longer length of glass is heated up, involves practically the same technique as followed in the cross-fire angle bend. The glass should be held in the correct manner and the required length rotated uniformly in the fire until it becomes plastic. The rotation then stops and the glass is moved from side to side until the proper

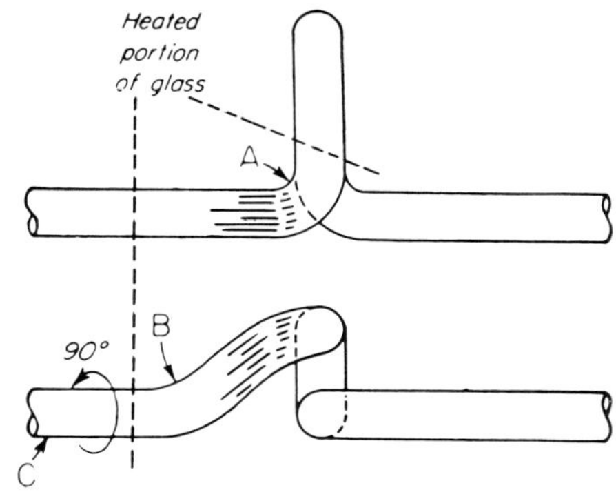

Figure 7.18. Combination angle bend. Note that the glass tubing at C is twisted inwardly 90 degrees after the two bends are formed.

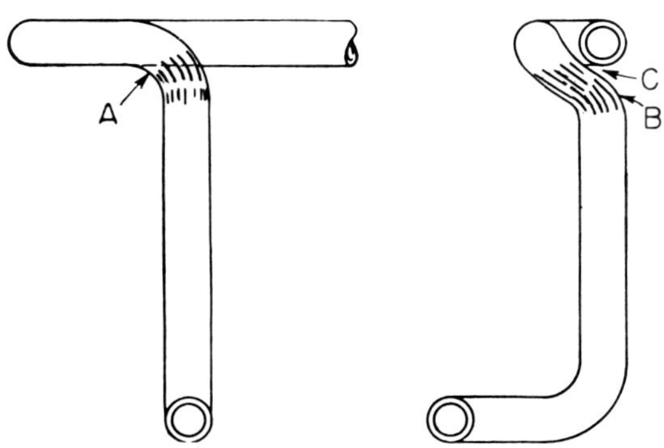

Figure 7.19. When making a combination angle bend for the letter T, the drop bend B must be long enough to clear the tubing C underneath the bend.

temperature is reached for the circular bend to be made. To illustrate the method of making a letter in the ribbon fire, the letter O, having an 8-inch diameter, is taken as an example. Figure 7.20(*a*) shows the letter O after all operations have been completed, including two double backs. Figure 7.20(*b*) shows a straight piece of glass. Figure 7.20(*c*) indicates the first bending operation of the letter O, which is a simple double back made in cross fires. Figure 7.20(*d*) shows the next operation completing one-half of the letter O in the ribbon fire. Figure 7.20(*e*) is the next step and almost completes the circular O. The next step is to make the second double back, as shown in Figure 7.20 (*f*), and then heat up the glass portion of the letter at this bend and form the cylinder contour of the asbestos pattern to complete the letter O, as indicated by the dotted line.

Before placing any portion of the glass in the fire, it should be chalk marked to indicate that portion which must be heated. It is also important that, when a circular bend is made, the straight portion of the glass is always brought out at a tangent to the circle on the asbestos layout, as shown on Figure 7.21. It is also important to allow the glass of a circular bend to cool before placing into the fire again. If the previous circular bend is still too hot, part of that bend will heat up faster than the cold portion and will cause a distortion in the circle.

The operator should make a preliminary study before going

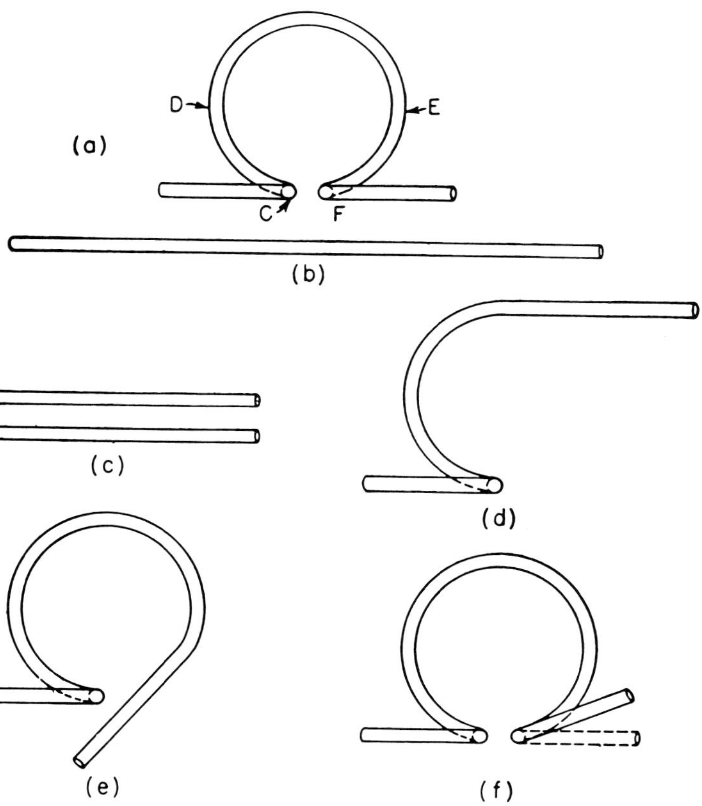

Figure 7.20. Sequence of operations required to make the letter O shown at (a).

ahead with his bending operation, especially if the work consists of a combination of circular bends, angle bends, double backs, and combination bends. The letter R is a good example because it incorporates almost all the bends which have been described.

The procedure in making the letter R is shown in the various stages in Figure 7.22. The first step involves making the circle of the letter R in the ribbon fire. Then all the various bending stages are indicated in the proper sequence.

This holds true for any other difficult letter having many different types of bends. Another good example is when a fairly

Figure 7.21. Matching a curved (ribbon-fire) bend to a layout.

complicated double-outline letter S is made from a single piece of glass. The various stages in bending are indicated in Figure 7.23. The first and important operation is to make the inside circular bend of the letter S, as shown in step (*b*). Careful study of the layout before bending should be followed when making a script or any other unusual or complicated job. The bending should always take place from left to right on the asbestos pattern.

Cutting and Splicing Glass. Besides being proficient in bending the glass, the operator should be capable of cutting and splicing the glass properly before he begins to make a tube. Cutting glass appears to be simple, but even this operation requires a certain knack. The glass, marked with chalk at the proper point, is scratched with a file or glass cutter's knife. The file mark should be made at right angles to the length of the tube, and it should be a definite cut into the glass about ⅛ inch long. The glass is then tapped gently with the file or knife at a point

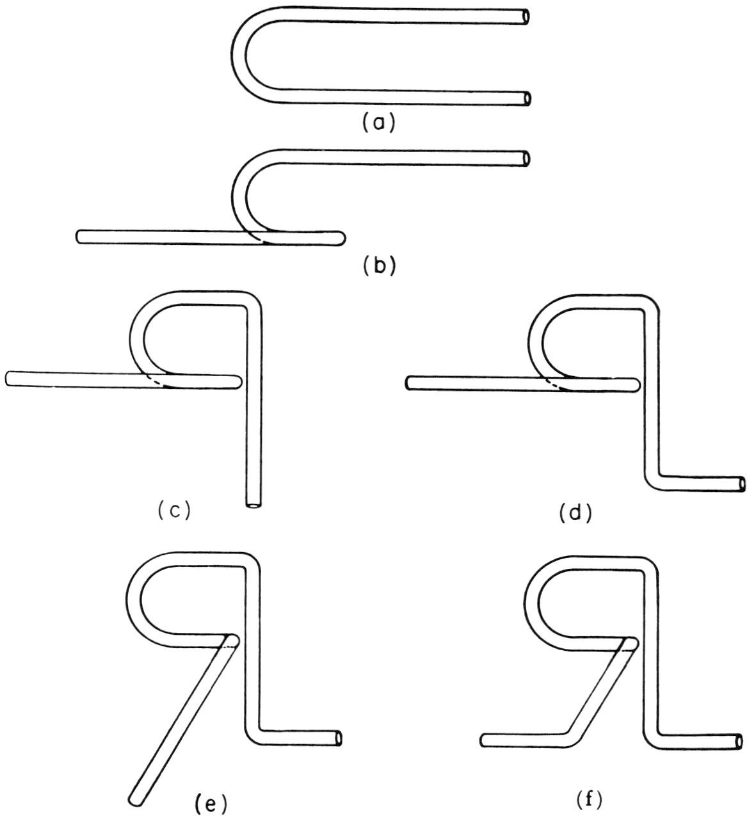

Figure 7.22. The various bending operations required to make the letter R.

exactly opposite the file cut, until it breaks. The break should be clean and even all around the circumference of the tube, as shown in Figure 7.24—not uneven, as shown in the second picture of the same figure. Poor glass or glass which is improperly annealed or not cool will tend to break unevenly.

A most useful tool for making the initial file cut in the glass is a flat hardened-steel file, 1/16 inch thick and ½ inch wide. The four edges of the file are used as cutting edges; as they become dull, they can be renewed by running the narrow edge of the file along an emery wheel or dry stone.

For delicate work or when an especially clean cut must be made, the following procedure is recommended: File a hairline surface cut into the glass at the position of the required cut and at the required angle, which usually is exactly perpendicular to the bore of the glass. Then heat the end of another piece of glass

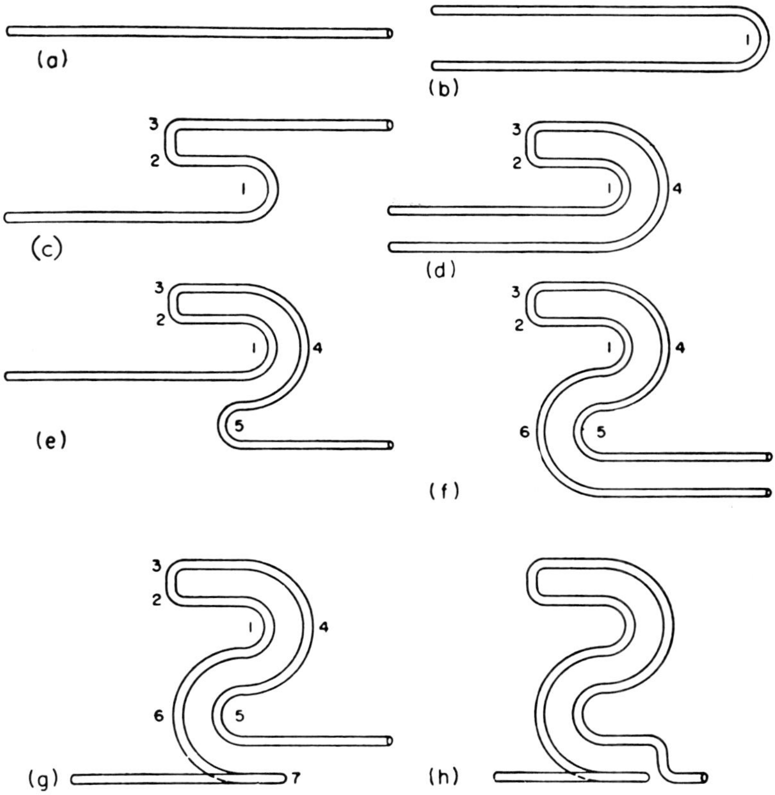

Figure 7.23. Sequence of bending operations required for a double-outline letter S in ribbon fire. The steps for the angle bends are not indicated.

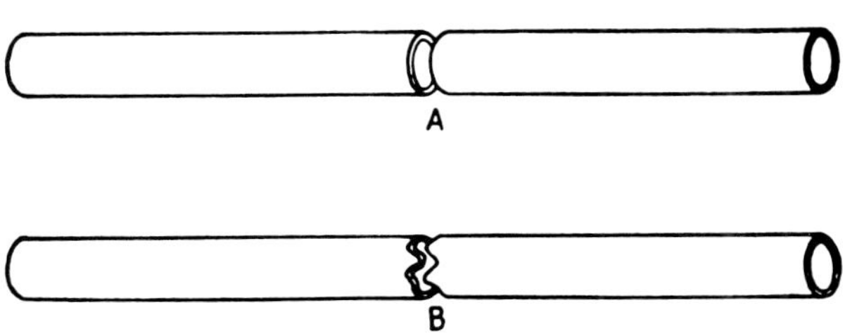

Figure 7.24. (A) A properly made cut in glass tubing. (B) A jagged cut, which cannot be used for a successful seal.

tubing until it has melted into a small molten bead of glass. Press this bead against the file mark in the tube to be cut. The glass will part at the cut in an even smooth break. The experienced operator should be able to cut glass time after time with such success that every piece cut can be stood on its end without falling or wobbling.

When two sections are spliced, it is necessary to cork one piece and to heat the two pieces of glass together in the flame either of the cross fire or the hand torch. In the cross fires, the glass is rotated; with a hand torch, the torch is rotated. When the glass has softened, the two edges are pushed together, and the open end of the tube is used for blowing the splice out to the proper wall thickness. The glass should be worked together and even reheated if necessary, to make sure that the two pieces are thoroughly fused together. Figure 7.25 shows finished letters being spliced to a straight tube in the cross fires.

Splicing Large Pieces. Splicing two sections of a tube together often involves large and unwieldly pieces of tubing which have already been formed into letters. When this is the case, the two sections should be laid on two asbestos-covered wooden blocks, on the worktable, so that the parts to be joined are firmly supported approximately one inch above the surface of the worktable. The hand torch is used to heat the two ends of the tubing to be spliced, as shown in Figure 7.26, the torch being slightly rotated and moved up and down between the glass ends, since the glass cannot be rotated in the flame. When the splice has been made but is not yet cold and hard, the blocks are removed, and the splice is laid over the proper part of the asbestos layout and made to match the outline of the layout as in the case of the angle or circular bend. A rubber-hose connection is used to provide a flexible means of blowing into the tubing. This tube should be about five feet long and is occasionally fitted at one end with a metal swivel joint, which prevents it from becoming twisted when the work is turned during the blowing process. It is important in splicing glass that only a small portion of each of the pieces to be joined be heated to the melting point. Usually, the glass should be heated not more than 1/16 inch back from the end. A joint usually requires two heating periods, one before and one after the joining of the two pieces.

Splicing Fluorescent Glass. A great deal of grief can be experienced in splicing two pieces of fluorescent glass if care is not taken in wiping away excessive fluorescent powder from the joint. The procedure is to wipe away about 1/16 inch of powder from the end of each piece of glass. This is especially important when splicing together glasses using calcium tungstate powders, such as blues or whites, as these powders, particularly in joints, give a great deal of trouble and may cause a leak. A joint that may seem to be airtight, even on the pumping system, may have an infinitesimal leak that will make the tubing fail after being installed in the sign.

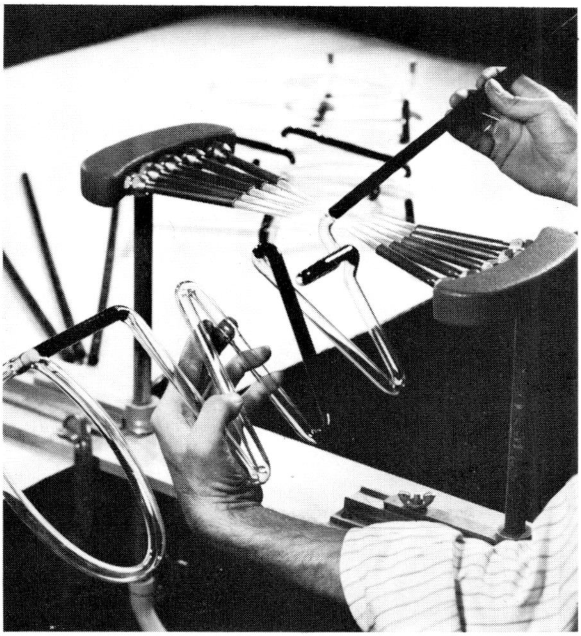

Figure 7.25. Splicing together two pieces of glass in cross fires.

Figure 7.26. Glass blower using hand torch to splice together two sections of tubing raised over asbestos layout.

Splices in exposed portions of the tubing, especially in fluorescent glass and in the middle of a sweeping curve, are conspicuous and should be avoided whenever possible. In large letters, however, such splices are often absolutely necessary, and in this case, special care must be taken to make a neat job, with a uniform wall and no lumps or uneven patches.

The Tubulation. The tubulation can be attached to a finished piece of tubing either by attaching it to the glass itself or to the end of the electrode. If it is attached to the glass itself, then it must be properly sealed by the glass blower. If the tubulation is attached to the electrode, then it is supplied complete (electrode and tubulation) by the manufacturer of electrodes.

The tubulation forms an important part of the unit because through it flow all the impurities that are released from the letter itself during the pumping operation. The tubulation should have a hole large enough to allow complete evacuation of the impurities in a unit in the quickest possible time. Many glass blowers constrict the tubulation near the unit for ease in sealing off the pumps. If the constriction is very small, the rate of flow is impeded considerably.

The tubulation is usually placed at the rear of the glass unit so that the seal-off will be hidden when it is mounted in place. The tubulation is attached as follows: A spot is chosen at the most convenient point on the unit, preferably near one of the electrodes. This point is heated with a hand torch, the heat being concentrated in a circle not larger than ¼ inch in diameter (see Figure 7.27). When the glass has become soft, the operator blows into one end of the letter, and the glass at the heated spot blows up in a small bubble about ⅛ inch high. Then the torch is applied again to the top of the bubble, and when it is soft, the operator again blows, bringing the bubble another small distance upward. The heat is applied the third time to the top of the bubble at a very small point and the bubble blown out. In this method of heating, a wall or ring of glass about ⅛ to 3/16 inch in height will form to which the tubulation can be attached with ease. It is spliced to the hole in the glass letter in the usual manner, the rubber hose being used for blowing, as in the splicing of two letters.

Many glass blowers do not use separate traps for their mercury but utilize the tubulation to the glass. After the tubulation has been attached, they heat a point of the tubulation glass about 4 inches from the unit. The glass is made very thick by pushing and gathering the glass at this point, and then it is blown to an uneven sphere, as illustrated in Figure 7.28, so that the larger portion of the sphere is on the bottom. The mercury can then be inserted into this trap without using a separate glass unit. However, extreme care must be taken with this type of mercury trap when the pump stopcock is first opened. Too sudden a withdrawal of air from the latter may suck the mercury into the manifold.

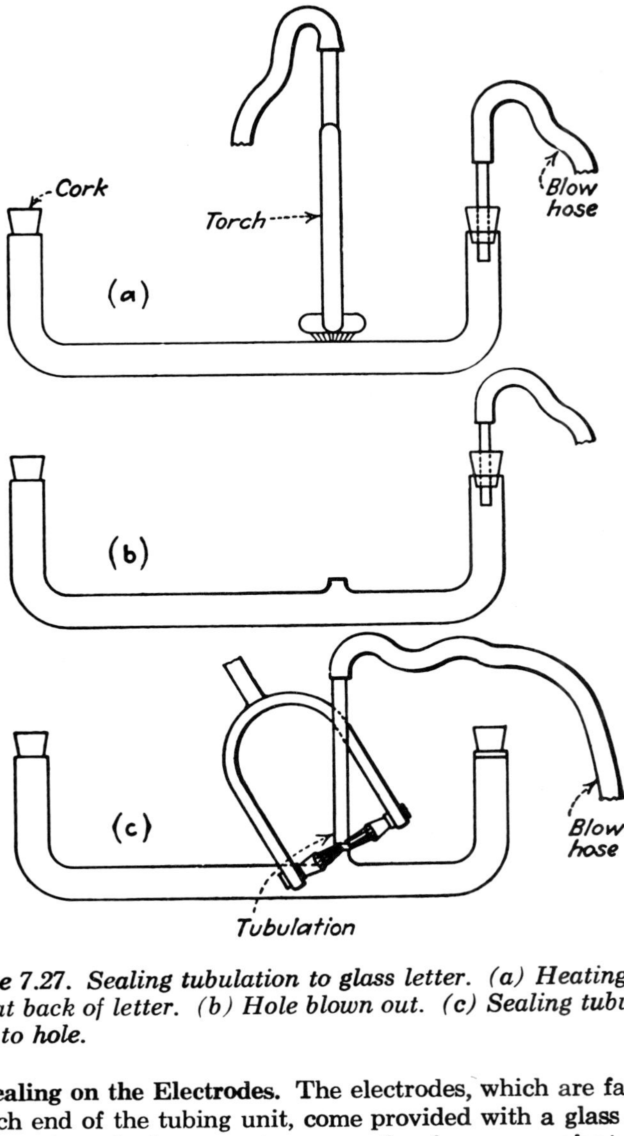

Figure 7.27. Sealing tubulation to glass letter. (a) Heating small area at back of letter. (b) Hole blown out. (c) Sealing tubulation glass to hole.

Sealing on the Electrodes. The electrodes, which are fastened at each end of the tubing unit, come provided with a glass jacket of approximately the correct cross section for a convenient splice. The larger electrodes may have a jacket whose diameter is too large for a good splice to the tubing unit. In this case, the jacket should be first reduced in size by heating the end to the melting point and allowing the glass to melt in. Once the electrode jacket and tubing have been brought to the same size, the splice is made in the same fashion as the ordinary splice between letters (see Figure 7.29). One end of the tube is corked, a rubber hose is attached to the tubulation, and one electrode is spliced. The cork is then removed, and the other electrode is spliced in place.

Working with Colored Glass. Data were supplied in Chapter 3 which indicated that the colored glasses have a higher annealing point than either of the clear glasses 001 or 002 or about 80° to 85°C higher. Special precautions must be taken to cool a colored glass slowly, so that it becomes properly annealed. Otherwise it will crack when subjected to strain. This is especially important when colored glass is joined to clear glass or to another type of colored glass. The general method of annealing, which

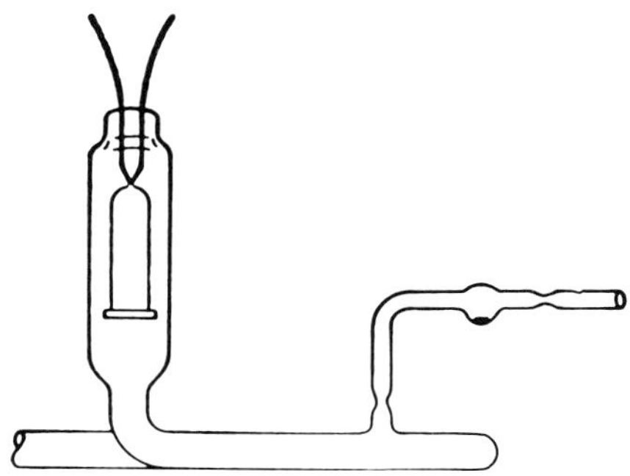

Figure 7.28. Method of making a mercury trap from tubulation glass.

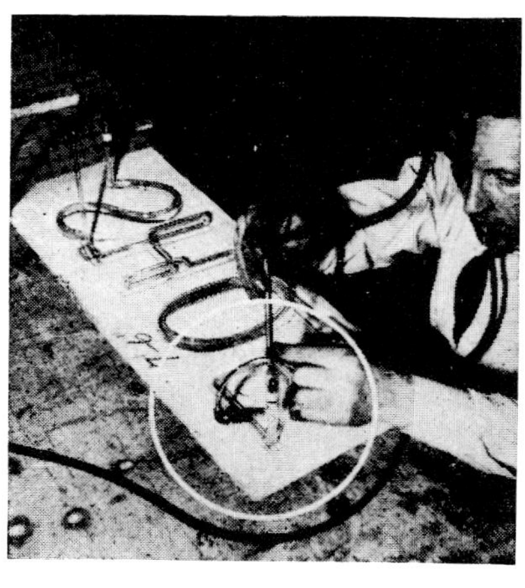

Figure 7.29. Sealing electrode to glass letter by hand torch.

should be followed when using colored glass, is called carbonizing and consists of holding the hot splice or bend in a flame from which all the air has been shut off. The gas of such a flame has a bright yellow color, is not nearly so hot as the blue flame, and will coat cool objects held in it with a black deposit of carbon soot. When the splice is first held in such a flame, it is too hot to take this soot deposit; but after it has cooled in the flame sufficiently, the black deposit will form, showing that the annealing is complete. The annealing process is simply a slow and gradual cooling of the glass which eliminates the strains which would be set up in the glass if it were cooled in the air directly.

Working with Pyrex Glass. Pyrex glass has a much higher melting point than lead glass and requires, therefore, a much hotter flame for successful work. It is true that some operators find that they can work pyrex in the ordinary air-gas flame used for lead glass, but the work is slow, tedious, and rarely if ever perfectly done. The reason is that pyrex will just barely soften in the ordinary flame; and while it can be worked in this state, it quickly hardens and has to be continually reheated.

The flame used for working pyrex is a combination of the usual illuminating gas and oxygen obtained from a tank of compressed oxygen. Special burners are usually required to mix the gases and handle the extra heat. The cannon-type burner is particularly useful for work with pyrex.

The pyrex type of glass handles differently from lead glass. It can be put directly into the intense blue oxygen flame without fear of cracking. When hot enough, it works easily, and it will not discolor under intense heat. Furthermore, it requires no annealing, and it is very strong when cold. When molten, it glows with a white glow, as contrasted with the red glow of the lead glass.

In general, the same procedure of heating and bending is followed with pyrex as with lead glass, with the exceptions noted in the preceding paragraph. Pyrex glass cannot be sealed with lead glass or in fact with any other kind of glass except pyrex.[1] If such a splice is attempted, it will be found that it is impossible to fuse the two glasses together. When it is necessary for some reason to join pyrex to lead glass, a graded seal should be purchased and used for the job. The seal is a length of glass tubing which contains seven different kinds of glass ranging from a lead glass at the one end to a pyrex at the other. The lead-glass end is sealed to the lead glass, and the pyrex end to the pyrex.

If the seal to be made between pyrex and lead glass is on the pumping system or elsewhere where a solid joint is not needed, a ground-glass seal may be used. This is simply a mechanically fitted joint between the lead glass and the pyrex, very similar in outward appearance to the ground-glass joint in a stopcock. The joint is filled with a special vacuum-system grease which keeps it vacuum tight.

Discoloration during Glass Bending. If lead glass is overheated in bending or splicing, it will become discolored with

[1] Nonex glass, one grade softer than pyrex, and hard uranium glass will seal to pyrex and are occasionally used. Hard uranium glass is often used for making a seal to tungsten leads.

a blackish deposit which forms in the glass itself. This color arises from the formation of free lead inside the glass, the lead being reduced from the lead oxide of the glass itself. The color may not be serious if it happens to occur in a place which will be painted out in the completed tube, but even in this case the glass is usually weakened. If the glass becomes discolored, it can be cleared up by heating the blackened part again in the fire, simultaneously reducing the gas fed to the flame and leaving the air at the same pressure. Since the free lead has increased the melting point of the glass at that point, considerable heat may be necessary to make it flow, but usually the black color can be made to disappear. The glass should be cooled slowly from this point.

Hints for the Glass Blower. Even the best glass blower is not perfect—and as explained in Chapter 3, glass has a way of being a very difficult material with which to deal when things begin to go wrong. As an aid in such cases, the following list of hints is given, hints which if considered will aid greatly in obtaining good results.

1. Insist on a proper air and gas supply and burners which can be adjusted to give the proper flame. No single item in the glass-blowing art is more important than a correct flame, correct in both shape and temperature. Also, air and gas supply must be constant at all times to prevent continual readjustment of fires.

2. Have the glass supply, electrode supply, tubulation glass, cutting tools, etc., close to the workbench. Walking for materials does not contribute to the bending of the tube, and it is an interruption which usually results in poorer work.

3. While bending the tube, keep in mind a mental picture of what the tube will look like when completed. This practice of seeing the bend in the mind's eye before making it will greatly aid the operator to do consistently good work without mistakes.

4. Guard against wasting glass. A good operator can save his employer much expense if he attempts to use glass economically. A glass waster is usually a poor glassworker.

5. If a splice or bend is bunched or lumpy, or if it is too thin, do not attempt to repair it. Cut back the glass and begin again; if absolutely necessary, begin the entire formation anew. This may not appear to be in line with the glass-conservation policy advised in the preceding paragraph, but a badly made splice can rarely be repaired no matter how much time is taken with it.

6. Do not attempt to make large curving bends too quickly. It is next to impossible to make a smooth curve if too much glass is heated at a time unless one is very experienced. Keep the unbent portion of the tube tangent to the curve on the layout.

7. Avoid blowing directly into glass tubing when making bends or splices. The bacteria from the mouth may discolor the

glass and become particularly harmful when blowing into fluorescent glass. Also, any fluorescent powder particles absorbed by mouth by the glass bender may be harmful to the health. *Use a mouthpiece and flexible rubber tubing together with an angle swivel in all stages of blowing and splicing the unit.*

CHAPTER VIII

Pumping Systems

Pumping System. The pumping system may well be called the heart of a luminous tube plant. The desired results of long life, high light output, and a minimum of discoloration of the finished tubing are virtually impossible if the exhaust system is not up to standard.

The primary object of the exhaust or pumping system is to remove all the undesired gas from inside the tube. Then, when the tube is filled with neon or whatever other rare gas is used, the gas in the tube will be pure and give maximum light of the desired color. If even a small amount of unwanted or impure gas is mixed with the rare gas, the color will be changed to a weak, bluish light. If a small amount of impure gas is present, the electrodes may be able to absorb this quantity without any bad aftereffects. However, if too large a quantity of impurities is present, the tubing may never be able to come up to color even after a long aging period.

The second object of pumping is to remove impurities from the walls of the glass and from the electrode shells. The high-current gas discharge set up during bombardment heats both the glass and the electrodes. This heat drives off the impurities from the walls and electrodes, in the form of gases and vapors. When the tube has become hot enough to fully release all impurities, the pump proceeds to remove the gases and vapors which have been liberated from the walls of the tube and from the electrodes. If these gases are not removed very quickly, that is, if the pump does not evacuate at a very high rate, the gases will be reabsorbed in the walls of the tube. Later, when the tube is installed and in operation, these reabsorbed gases may be liberated and may destroy the brilliance and color of the tube's light. But if the pumping is rapid, the gases released by bombardment will be drawn out before they have had a chance to collect again on the walls of the tube. After bombardment, the tube must be pumped out to a very high vacuum, that is, to a few millionths of atmospheric pressure. The rare gas (or rare gas and mercury) is then inserted, and the tube is sealed off the pumping system. The luminous tube is then complete, ready to be mounted on the sign.

The Requirements for the Pumping System. From the foregoing it can be seen that the following requirements of a good pumping system must be met if good luminous tubing is to be produced:

1. The pumping system must be rugged and vacuumtight. The stopcocks used throughout the manifold and for the rare-gas containers should be of the high-vacuum type. They also must be specially treated with grease to make them free from even the slightest leak. Poorly ground stopcocks, no matter how well greased, will have a tendency to leak.

2. The pumping system must be clean. Dirt, water vapor, or excess stopcock grease in the system will greatly impair the vacuum.

3. The pumping system should be capable of producing a vacuum better than one micron (about one-millionth of atmospheric pressure—see sections below, which explain the nature and meaning of vacuum measurement). This vacuum is much higher than that used in electric light bulbs and about the same as that required in radio tubes. Many old mechanical pumps used by luminous tube manufacturers today cannot alone produce the necessary vacuum. Either a new pump should be substituted, or the mechanical pump should be augmented with a diffusion pump placed in series with the mechanical pump.

4. The pumping system should exhaust or pull out at a fast rate. After the tube has been bombarded, it should pull down to a vacuum of two microns in less than two minutes. If the system cannot do this, the gases freed during bombardment will be reabsorbed before they are pulled out.

5. The entire system should be reliable. A rough-and-ready system which will work regardless of the temperature or humidity and regardless of a certain amount of abuse is the only practical unit for a neon plant. Beyond occasional replacement of the oil in the mechanical pump and occasional replacement of the glass in the manifold, the pumps should require no servicing.

The Nature of a Vacuum. Before we go on to describe the various parts of the pumping system and their operation, it is necessary first to consider what a vacuum is, how it is measured, and how it can be obtained.

Three hundred years ago, man did not know how to obtain a vacuum. He may have imagined a space completely empty of every material thing, but he had never succeeded in actually producing a vacuum. Torricelli, the famous pupil of Galileo, discovered that if a glass tube, more than 30 inches long and closed at one end, were filled with mercury and then inverted into a glass also filled with mercury, the mercury in the tube would be sustained to a height of not more than 30 inches. The space above the mercury in the tube was supposedly a complete vacuum. We know now that in this space there is actually a quantity of mercury vapor, but at that time the existence of the vapor was not suspected.

The mercury is sustained to the height of 30 inches by the pressure of the air on the mercury in the cup. As the air pressure varies from day to day, the height of the mercury in the tube varies also. Such variations in atmospheric pressure are indications of changes in weather. The mercury tube thus became the first barometer.

Vacuum Measurement. The relation between the air pressure and the height of the mercury column has led to the use of the barometer as a measurer of pressures. In fact, most pressure gauges used in vacuum measurement today work on the barometer principle. It was found that the average pressure of the air at sea level was capable of sustaining a column of mercury to a height of 760 millimeters—29.92 inches—which is the accepted standard of air pressure. In other words, it was convenient to express pressures in terms of the length (in millimeters) of the mercury column which that pressure would support. So today we find pressure measured in *millimeters of mercury,* abbreviated as "mm Hg."

A complete vacuum is a completely empty space, that is, within which not a single molecule of gas or other matter can be found. No perfect vacuum exists on earth, and even in the empty spaces between the stars there are at least 28,000 molecules of matter in every cubic foot of space. The best vacuum ever obtained on earth by man is one in which there were 100,000,000 molecules in every cubic centimeter, and yet this apparently poor vacuum is equal to 0.000,000,000,13 of atmospheric pressure.

Even with simple pumping systems, pressures of less than a millionth of atmospheric pressure can be obtained. Since such pressures would sustain a column of mercury to a height of only 0.007 millimeter, a more convenient unit for measuring vacuums, the *micron,* has been adopted. The micron is a thousandth of a millimeter. The required vacuum for luminous sign tubing before filling is about one to five microns and for cold cathode tubing about 0.1 micron.

Relation between Pressure and Vacuum. The reason why the pressure inside a container is tied up with the degree of vacuum in the container is not always understood. A container of gas is in reality full of millions of molecules of gas. These molecules are flying about in every direction at very high speed. As the molecules near the wall of the container hit the wall, they exert a pressure on it. In other words, the pressure is really caused by a continual battering of moving molecules. Now, as the container is pumped out, fewer and fewer molecules remain; consequently there are fewer molecules available to batter the wall of the container. Hence, as the vacuum increases, the pressure goes down. Thus the pressure can be used to measure the vacuum.

Pressure of Rare Gas. After the tube has been bombarded and thoroughly pumped, it is filled with the rare gas (neon, argon, etc.). The pressure of this gas is of the greatest importance. If the pressure is too low (one millimeter or less), the tendency for the electrodes to sputter will be great. If the pressure is too high, the light output will tend to decrease and the resistance of the positive column will increase. So the pressure of gas inserted in the tube must be accurately measured.

The same sort of gauge used for measuring the vacuum can be used to measure the gas pressure, except that the gas pressure gauge must be able to read much higher pressure. The vacuum gauge reads microns, or thousandths of millimeters. The gas pressure gauge is calibrated in millimeters. Pressures of 6 to 20 millimeters are usually used for luminous tube sign tubing. For cold cathode tubing, especially of large diameter, gas pressures as low as four to six millimeters are used.

One further fact about vacuums is of importance. When liquids are present in an evacuated vessel, they tend to evaporate and fill the space of the container with a vapor. The pump will draw out these vapors along with the gases in the tube; but as long as any liquid remains in the tube, the vapor will be continually renewed. Hence the entire vacuum system should be free of liquids. If liquids, greases, or oils must be used, they should be of specially chosen types of low vapor pressure, so that the amount of vapor that they give off in the vacuum system is small.

The Parts of the Pumping System. Pumping systems used by luminous tube manufacturers vary according to the amount of work handled. Some systems are elaborate, others very simple. But every pumping system must contain the following essentials: the pump itself, either a single mechanical pump or a mechanical pump and a diffusion pump in series; the pump manifold, with stopcocks; a vacuum gauge which measures the ultimate vacuum; a millimeter gauge for measuring gas pressure; and several containers of the rare gases which are connected to the manifold through stopcocks. Some of the more modern systems use two pumps. The first, the "fore pump," is used for getting a rough vacuum. Then, by means of suitable stopcocks, the tubing is switched over to the high-vacuum pumping system which usually consists of a good mechanical pump in series with a diffusion pump.

Plant Layout. The layout of the pumping system should be as compact and neat as possible. The one shown in Figure 8.1 is a typical example, but this particular style may not be adapted to every plant. Figure 8.2 shows the parts of a system laid out on a flat plane. Many pumping systems have been built haphazardly; gas containers and extra manifold sections have been added as the need arose. When a system has been built in this manner, it seldom presents a good appearance, and it is seldom as efficient as it might be. The pumping system should be built large enough in the first place to take care of the increase of work which can be expected as the business grows. The original layout should

Figure 8.1. A typical pumping table. 1. Mechanical vacuum pump with motor. 2. Rubber vacuum hose. 3. Glass tubing. 4. Diffusion pump. 5. Graded glass seal. 6. Glass manifold. 7. Rare-gas containers. 8. Gas stopcocks. 9. E-Z Read gas-pressure gauge. 10. Bombarder. 10A. Metal case covering bombarder. 11. Bombarder high-voltage leads. 12. Battery clips with rubber hoods. 13. Asbestos covering on top of table. 14. Low-voltage switch box. 15. Magnetic-switch contactor for bombarder. 16. Regulator for bombarder. 17. BX cable. 18. Push-button switch for operating contactor. 19. Electric switch for pump. 20. Terminal socket. 21. Red lamp indicator for bombarder. 22. Mercury U gauge. 23. Main stopcock. 24. Blow-hose stopcock. 25. Glass connection to tubing to be pumped. 26. Testing coil. 27. Table. 28. Wood frame for gas containers and manifold.

be given care and thought before it is actually installed. A neat layout is not a luxury. If the pump man works on a neat system, he will do better work. Repairs can be made more quickly, leaks are easier to find, and the whole system can be kept in order with less bother. Of course, a pumping system which resembles that of a college laboratory is neither necessary or desirable; but a good sensible layout is almost as important as are good pumps.

The Manifold System. The pumping manifold is the part of the pumping system to which the tubes are attached for pumping.

The manifold is a system of glass pipes arranged so that one or more tubes can be attached to it and having an assortment of stopcocks for use in connecting the pumps and the rare-gas flasks at the proper time. The glass used in the manifold system is usually the same as that used for the luminous tubes. In practically all cases, lead glass is used. However, systems can be constructed of pyrex glass which can be connected to lead glass by suitable graded seals. The glass tubing used for manifolds should be large—15 to 18 millimeters in diameter—except where the tubulation is to be joined to the system. At this point the tubing should be of the same size as the tubulation.

The manifold should be large enough to handle the work which will be put on it, but it should not be too long. If it is long, it will take a long time to pump out, and in addition it will waste the rare gas which must fill it when the tube is filled.

All splices in the glass manifold must be properly sealed. Thin seals cause breakage due to vibration, and heavy seals cause cracks due to strain and stress set up in the glass. All glass should be well annealed.

A manifold layout, showing gas containers, gauges, all necessary stopcocks, pump connections, and tubulation glass for sealing on the tube, is shown in the manifold section in Figure 8.2. The rubber blowtube and stopcock shown on the manifold are used to permit the glass blower to blow into the system while sealing on the tube to be pumped and for the purpose of admitting air to the system whenever this may be necessary.

On some systems, special drying traps are used, which contain phosphorus pentoxide, a chemical which will absorb water vapor; if such a trap is used, the phosphorus pentoxide should be replaced regularly; otherwise it will become saturated with water vapor and do more harm than good.

Stopcocks. The stopcocks used throughout the system should be of standard design. They may be bought from glass-supply houses and are fitted with lengths of tubing at each end, which must be spliced to the rest of the system. The stopcock must have a carefully made ground joint, between plug and bore. It is so constructed that when it is open, as shown in Figure 8.3, the handle is parallel to the length of the tube. A glass stopcock used for a vacuum system should have a more carefully ground fitting than the standard chemical stopcocks available on the market. Even a good ground-glass fit is not sufficient unless specially treated with a suitable stopcock grease, specially made for vacuum work. This grease is smeared over the ground-glass surface of the inner portion of the stopcock in a thin coat, but care must be taken not to get grease in or near the hole through the stopcock, since this would clog the opening and allow the grease to get into the manifold. The inner portion (plug) of the stopcock, after being covered thinly with grease, is turned into the outside portion (or seat). When properly greased, no air

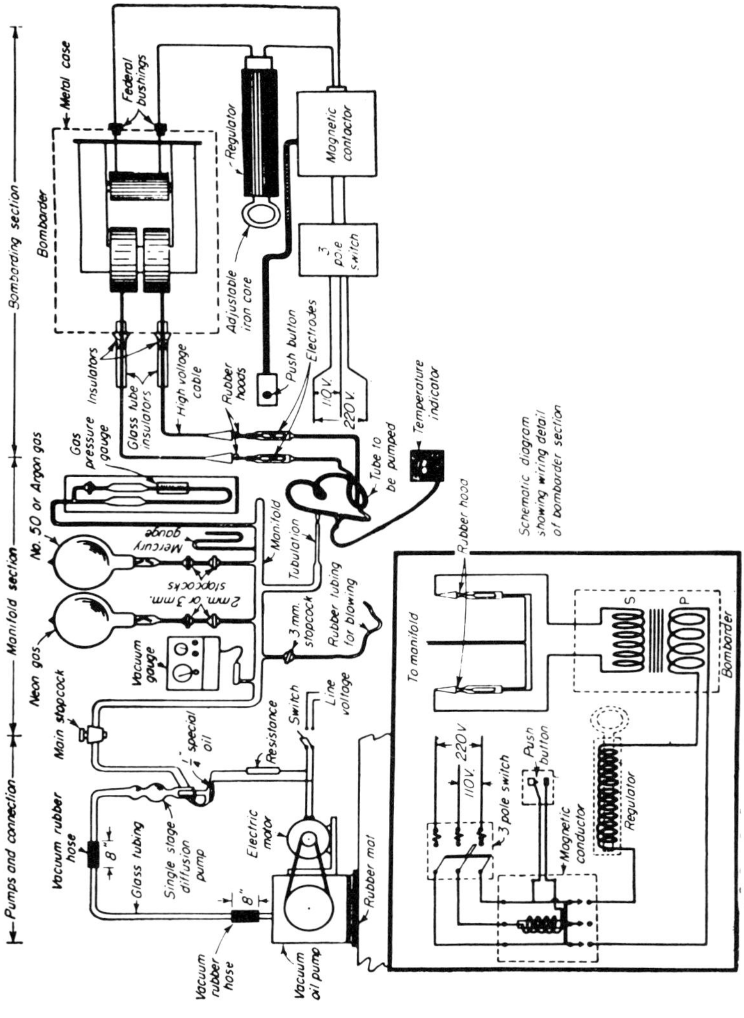

Figure 8.2. Schematic layout of a complete pumping position. At the lower left is shown the detailed wiring diagram of the bombarder section.

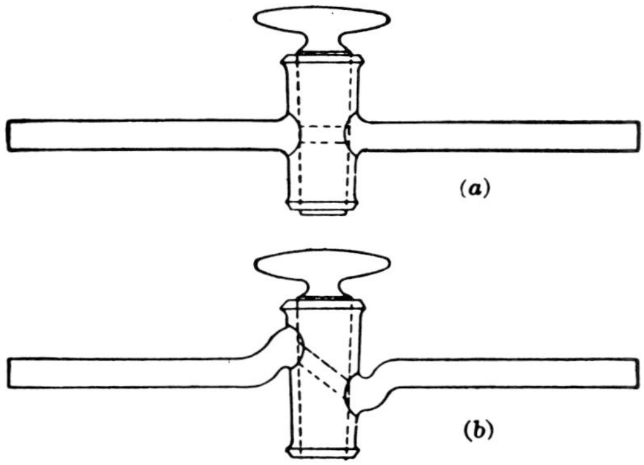

Figure 8.3. Two types of stockcocks: (a) Straight bore. (b) Oblique bore.

bubbles will be seen between the two ground-glass surfaces, and the stopcock should turn easily and should be completely air-tight. Grease used for stopcocks should have a very low vapor pressure; only the finest quality should be used, since poor-quality grease has a tendency to release vapor within the pumping system.

Stopcocks are available in many different types. The standard types shown in Figure 8.3 can be obtained with either a straight or an oblique bore in the plug. Figure 8.4 indicates one type

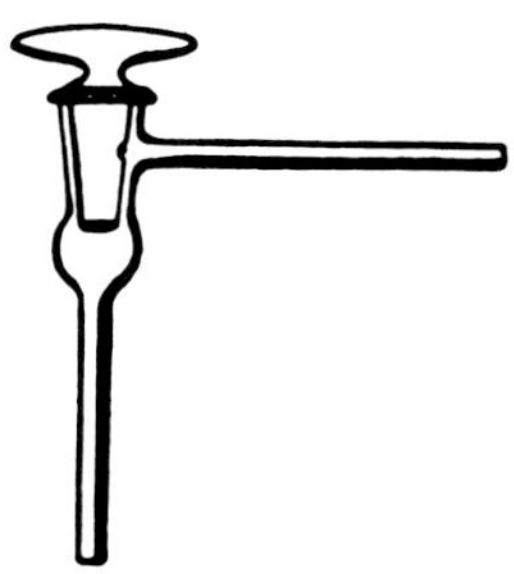

Figure 8.4. A hollow-bore stopcock.

of hollow-bore stopcock which can also be used, principally for the main stopcock in a pumping system. Other stopcocks are available such as mercury-filled types, but these are not practical for a rugged luminous tube system. Generally, long-bore stopcocks with two, three, or four millimeter openings in the plug are used for the rare-gas flasks, for connecting to the glass-blower's blow hose, for the gas-pressure gauge, and for any other needs in the system, except for the main stopcock. The main stopcock to the pump should have a plug with at least an eight to ten millimeter hole; otherwise it may interfere with the pumping speed. Stopcocks have to be regreased since the grease coat wears out after a time. Many hard-to-find leaks can be traced to poorly greased stopcocks or to stopcocks that have a trace of foreign matter between the plug and the bore.

The Rare-gas Containers. The rare-gas containers must be attached to the system very carefully so that none of the gas is wasted and so that the gas inside the container remains pure. The method is outlined in Chapter 9.

Some systems have only one stopcock on the rare-gas flasks. This simple installation will serve satisfactorily, but the double-stopcock installation is much safer, will conserve the gas, and will permit more accurate filling of the tube. After the stopcock nearest to the gas flask is opened, it permits the filling of the space between the two stopcocks. The first stopcock is then closed and the second stopcock nearest to the manifold opened and the gas "ladled" into the tube. This ladling process can be duplicated as many times as needed to reach the proper gas pressure in the luminous tube.

Mechanical Pumps. The mechanical motor-driven pump can be used in either of two ways in the luminous tube pumping system. It may be the only pump on the system. In this case, it must do the entire work of pumping, and for this reason it must be a very good pump. Or it may be used with a diffusion-type pump. In this case, the mechanical pump is used simply to provide a rough vacuum on the output side of the diffusion pump, while the diffusion pump does the high-vacuum pumping. In this case, the mechanical pump need not be capable of drawing so high a vacuum as when it is used alone. But it must be fast whether used with a diffusion pump or not. Regardless of its intended use, the mechanical pump should be carefully made to withstand long and hard usage.

The mechanical pump has two connections: the intake port, which is connected to the manifold or diffusion pump; and the output port, which forces the pumped gases into the air. The pump can be connected to the motor by a direct drive, chain drive, or belt drive. The motor should have the proper horsepower rating recommended by the pump manufacturer, and it should be firmly mounted on the same base as the pump itself.

The mechanical pump is filled with a special oil having a low water content, which seals its valves and pumping action. This oil should be replaced occasionally, and it should never be allowed to fall below the level indicated on the inside of the pump casing. The oil eventually becomes contaminated with water vapor and other impurities after about six months of use and should be replaced at the end of that time.

Rubber hose is usually used to connect the intake port of the pump to the glass tube which leads to the diffusion pump or manifold system. This hose should be of pure gum rubber and of heavy wall construction; it should be bound to the intake port with wire. Sealing the rubber hose to the pump intake and to the glass should be done with castor oil or some other lubricant which will make the joint airtight.

Almost all modern high speed vacuum pumps, like those in Figure 8.7, work on the principle shown in Figure 8.5 (*a*). The

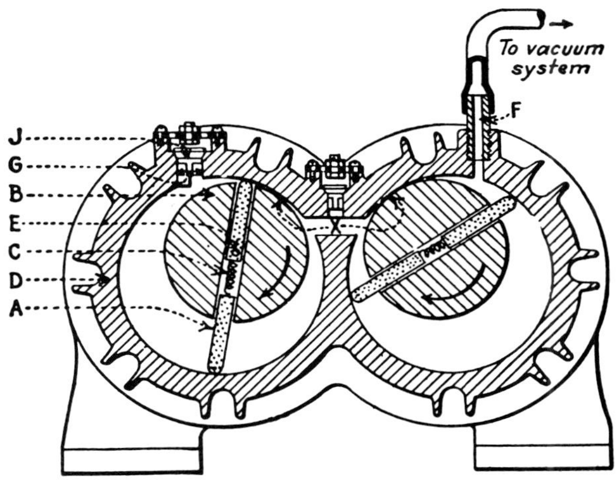

Figure 8.5(a). Compound mechanical oil pump with rotary sliding vane. A. Sliding vane. B. Rotor. C. Spring to force vane against cylinder wall. D. Cylinder or stator. F. Intake port. G. Outlet port. J. Outlet valve. X. Rotor clearance.

pulley drives a shaft on which a heavy vane slides, so that the edges of the vane are always in contact with the edge of the circular pump chamber. The drive shaft is off the center of the chamber as shown, so that the vane moves eccentrically. The vane, sealed against the side wall by the pump oil, sweeps the air admitted by the intake port around to the discharge port, where it escapes. This action is continually repeated as the pump revolves. Almost all pumps are really two pumps in one (compound pumps), connected in series and operated so that the two vanes turn together. This series arrangement adds greatly to the pumping speed and permits much higher vacuums to be attained.

The sweeping-vane pump has been modified in the eccentric-cylinder type [see Figure 8.5(*b*)]. The sweeping vane has been replaced by a cylinder mounted off center on the drive shaft. The cylinder rests against a sliding stationary vane which is sealed in the cylinder by the pump oil. As the cylinder rotates, it sweeps the gas from the intake port to the discharge port in much

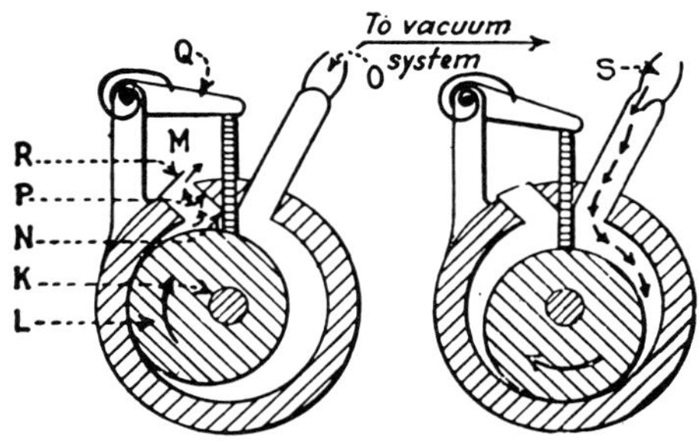

Figure 8.5(b). Cenco pump, showing rotor in two different positions. K. Main shaft. L. Rotor. N. Scraping vane (fixed but movable vertically). O. Intake port. Q. Compression lever which forces vane against rotor. R. Outlet port. This pump is compound and has two units in series.

the same manner as that of the sweeping-vane pump. In this pump, the sliding action is moved from the drive shaft to the side of the pump wall, and as a result the pump is easier to construct and maintain in good working order.

Other types of pumps are occasionally used, one of which works on a rotating-gear principle, the teeth of the interleaved gears being used to sweep the air from intake to output, but these pumps wear out quickly after a few months of service. In addition to the gear pump, there are piston pumps very similar in action to the piston hand pump used for filling automobile tires but with the action reversed. Piston pumps are not fast, and they cannot produce a high vacuum; so they are rarely used.

The most effective type of mechanical pump is the so-called molecular pump (see Figure 8.6). This works with a reduced

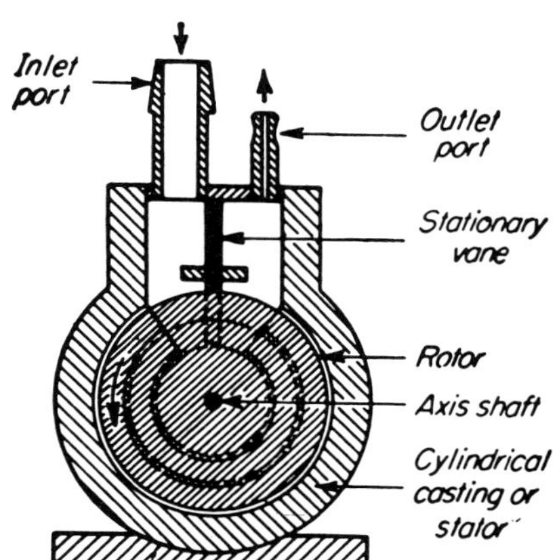

Figure 8.6. Section of a molecular-type vacuum pump.

pressure vacuum on its output side, provided by an auxiliary "backing" pump. The molecular pump works on the principle of molecular drag; that is, a rotating cylinder or rotor operating at a very high speed (4,000 to 10,000 rpm) is used to exert a dragging force on the gases from the intake to the output. When properly operated, molecular pumps can give vacuums of 0.001 micron when provided with a backing pressure of not more than ten microns. But they are expensive and delicate pumps, and they are not used in the luminous tube trade.

The cubic content of a luminous tube is relatively small. A cold cathode tube 25 millimeters in diameter by eight feet long has a cubic content of less than one liter. To this, of course, must be added the capacity of the manifold of the pumping system itself, but the total will not average more than one liter. Because of this small volume which has to be evacuated, a relatively small mechanical pump can be used. It must be able to pull out the gases which are released during bombardment at a high enough rate to prevent the gases from being reabsorbed. Pumps available on the market today are the Tubelite VOP-5, the Cenco Hyvac, Megavac, and Hypervac 23. Each has a free air displacement in liters per minute sufficient to cope with any pumping problem which may be encountered in the luminous tube industry. The Hyvac vacuum pump has a free air displacement of approximately ten liters per minute; the VOP-5, approximately 28 liters per minute, the Megavac at 325 rpm has a free air displacement of 31 liters per minute; and the Hypervac 23, 240 liters per minute. At a vacuum of one micron, the speed of pumping of liters per second is reduced to approximately 1/125 that of the free-air displacement.

In general, a satisfactory pump for the average small luminous tube shop has a displacement of 15 to 40 liters per minute. When a large installation is to be made and a number of cold cathode tubes are to be evacuated at the same time, a pump with a higher free-air displacement can be used.

When a mechanical pump is turned off, the oil within the pump may be sucked up into the manifold or diffusion pump unless special precautions are taken. Some mechanical pumps are fitted with a nonreturn valve which prevents the oil from backing up out of the pump. Other pumps do not have this arrangement. No matter what kind of pump is used, it is always best to allow air to enter it from the manifold after it is stopped. If a diffusion pump is used, then the air should not be let into the system until the diffusion pump has cooled sufficiently to prevent oxidation of the pump liquid.

Occasionally even this precaution is of no avail, and the pump oil does get into the diffusion pump and manifold. When this happens, there is nothing to do but take down the entire system; clean it out with lye, carbon tetrachloride, or some other grease

Figure 8.7. Three types of standard mechanical vacuum oil pumps used in the neon sign industry. (a) Tubelite pump. (b) Cenco pump. (c) International pump.

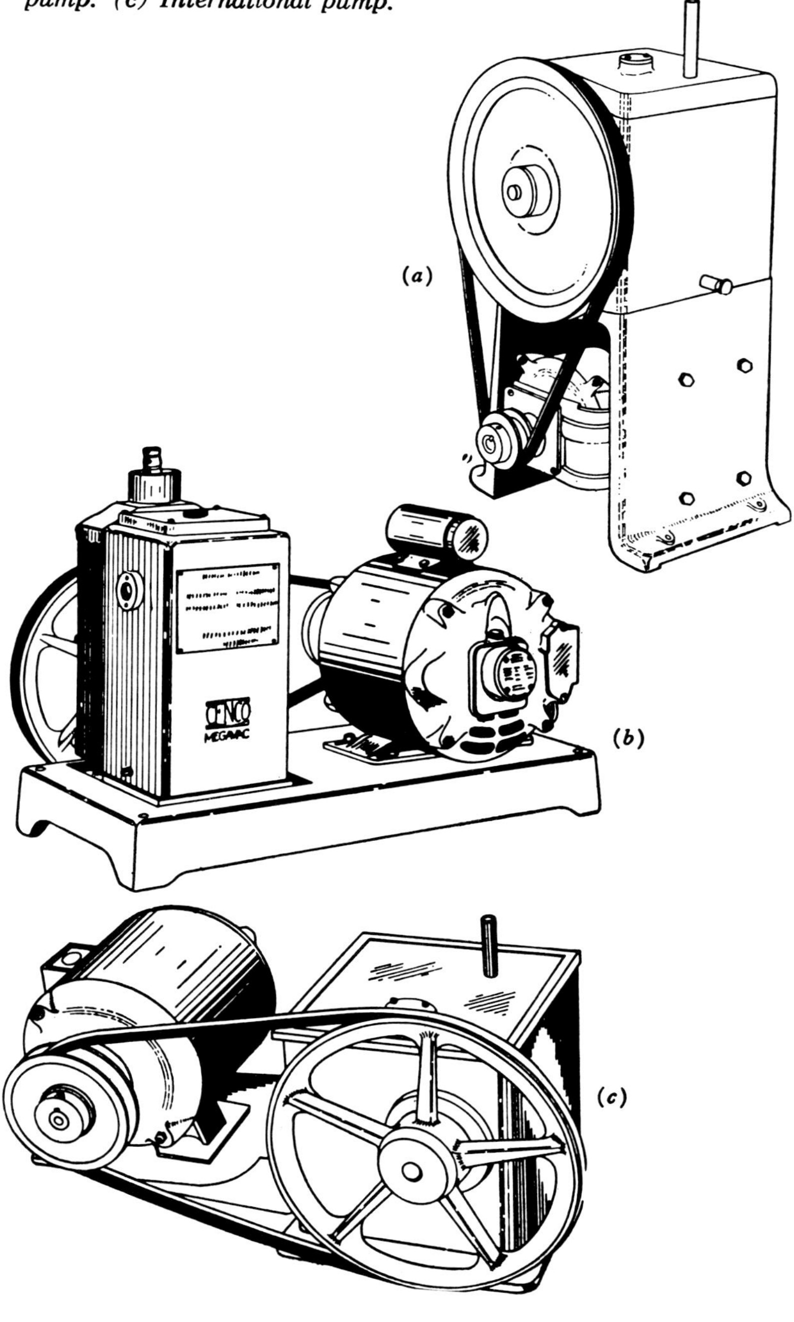

remover; refill the diffusion pump with mercury or organic liquid; and seal the system.

Table 8.1 gives the pumping speed and highest attainable vacuum of two of the better known kinds of pumps.

Diffusion Pumps. Very few luminous tube manufacturing plants depend upon a mechanical pump alone. Almost every manufacturer has come to realize that some sort of diffusion pump is necessary for a high speed pumping system. The reasons for the superiority of the diffusion pump are many: It is independent of humidity changes, pumping equally well in wet and dry weather. It is extremely fast. It does not wear out, and it requires a back pressure which can be supplied by any fast mechanical pump.

Table 8-1. Comparative Pumping Speeds

Type of glass pump	4,000-cc volume under laboratory conditions, minutes*		Estimate for 275-cc average tubing units under plant conditions, minutes†	
	Hyvac fore pump	Megavac fore pump	Hyvac fore pump	Megavac fore pump
No diffusion pump	33	20	7	4
Single-stage water-cooled mercury pump	18	16	Less than 4	Less than 3
Single-stage butyl phthalate pump	20	..	Less than 4	Less than 3
Three-stage water-cooled mercury pump	9	7	3	2

* The figures given represent the time required to pump out the volume indicated to a limiting pressure of ½ micron. P. J. Mills, "Time Pressure Characteristics of Various Diffusion and Molecular Pumps," *Review of Scientific Instruments*, Vol. 3, No. 6, June, 1932.

† Authors' estimate.

Three types of diffusion agents are commonly used: (1) mercury, (2) organic liquids, and (3) silicones. The latter types have almost entirely displaced the mercury type. These pumps can be constructed either of glass or metal and can be either air-cooled or water-cooled, depending upon the amount of volume which the pump handles.

The metal pumps are generally of the single-stage type. The glass type can be constructed in many different ways and can be single jet, single compartment; double jet, double compartment; double jet, triple compartment; or triple jet, four compartment. For the luminous tube industry, either the single-jet, single-compartment, or double-jet, double-compartment pumps are generally employed.

The simple diffusion pump consists of a glass reservoir containing the liquid, a heater which heats the liquid, a cooling chamber in which the vapor from the liquid is recondensed by means of a special cooling system, and a tube for conducting the recondensed vapor back into the main liquid reservoir.

Mercury was employed as a pumping liquid first. The uncertainty of performance with this relatively high-vapor-pressure fluid made its use undesirable since it is easily affected by slight contamination. This has resulted in widespread use of organic liquids and silicones.

The organic liquids were developed by the Eastman Kodak Co. research laboratories in 1930 when butyl phthalate was first used as a pumping fluid. Butyl phthalate is still commonly used for pumps designed for the luminous tube industry. Other liquids have since been developed for diffusion pumps such as Octoil, Octoil-S, Amoil, and Amoil-S. Neither Octoil nor Octoil-S is recommended for the luminous tube industry because the oil readily becomes oxidized as air rushes in at the beginning of a pumping cycle. An Octoil diffusion pump is generally used with its own mechanical pump only when a separate fore pump is employed to obtain a rough vacuum first.

Silicone fluids have also been developed for use with diffusion pumps. The silicone fluids are highly recommended because of their inherent thermal stability, which permits a high-vacuum system to be open to air without decomposition or reduction of pumping speed. However, the silicone fluids are relatively expensive.

To illustrate the operation of a diffusion pump, the layout of a single-stage diffusion pump is shown as an example in Figure 8.8.

In the diffusion pump, the pool of oil or silicone liquid is heated by a heater, and the vapor which forms above the pool is drawn toward the intake part of the mechanical pump. But the vapor chamber is also connected to the glass manifold, and as the cloud of vapor is drawn toward the mechanical pump it sweeps before it a portion of gas from the manifold. The vapor is the pumping agent; it half drags, half pushes before it the gases from the manifold into the mechanical pump.

If the vapor was not collected, it would pass out into the mechanical pump and be lost with the impurities, and eventually the supply of liquid in the pool would be completely evaporated. So some means of collecting the vapor must be provided. This takes the form of a condensation chamber which cools the vapor. By the time the vapor has reached the chamber, this cooling effect has condensed it into tiny droplets which collect in the side tube and roll back into the main reservoir. By the time the condensation takes place, a fresh supply of vapor has formed and is pushing more gas from the manifold into the mechanical pump, so that the action is continuous as

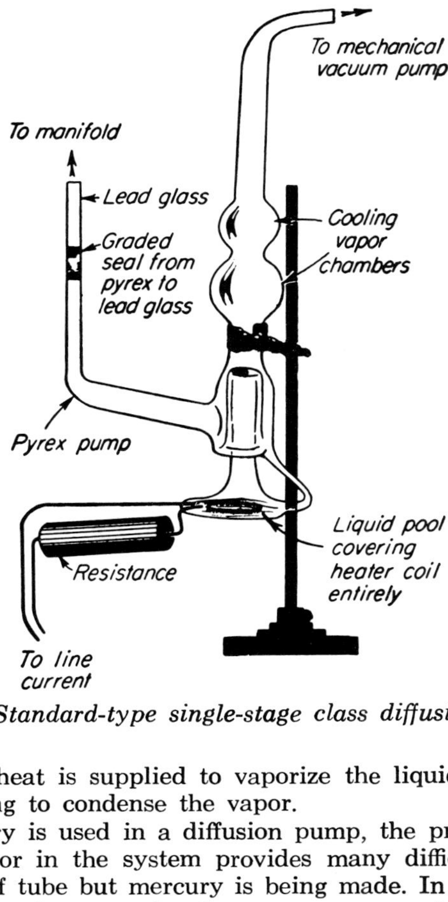

Figure 8.8. Standard-type single-stage class diffusion pump.

long as the heat is supplied to vaporize the liquid and the required cooling to condense the vapor.

If mercury is used in a diffusion pump, the presence of the mercury vapor in the system provides many difficulties if any other kind of tube but mercury is being made. In the case of a neon gas tube, for example, the mercury vapor from the diffusion pump may get into the tube and cause a blue glow where only red is wanted. So a specially devised vapor trap must be provided between the diffusion pump and the manifold (see Figure 8.9). This trap is a tube surrounded by a very cold substance, such as liquid air or chopped-up dry ice. The intense cold at this tube condenses any mercury vapor and traps it before it can reach the manifold system.

The Effect of Temperature on the Operation of the Diffusion Pump. It is of the greatest importance to maintain an even temperature in the boiler containing the liquid, that is, mercury, organic liquid, or silicone fluid. If the temperature is too high, violent boiling may cause the liquid to be vaporized to such an extent that it may be pulled out by the mechanical pump. The organic liquids may become crystallized under intense heat. For this reason, the temperature control in the liquid chamber is of prime importance. Too high a temperature also impairs the effectiveness of the pumping action, since the greatest amount of gas will be drawn from the manifold only if the vapor pressure is correct—neither too high nor too low. If the temperature is too low, not enough vapor will be produced to create a heavy drag on the gas in the manifold, and the pumping will be slow.

The diffusion pump can be cooled by water circulation, by forced air, or by having a large enough cooling chamber to allow for proper condensation by cooling from the surrounding atmosphere.

Organic-liquid or silicone-filled pumps require a lower backing pressure than the mercury type of pump; they operate at maximum speed when the backing pressure is about 0.01 millimeter of mercury. Almost any good mechanical pump is capable of providing this back pressure for the pump, so that manufacturers need have no fear about replacing their mercury pumps with an oil pump. In general, it may be said that the mercury and oil-condensation pumps will perform equally well for use in tubing plants, but the greater convenience of the oil type

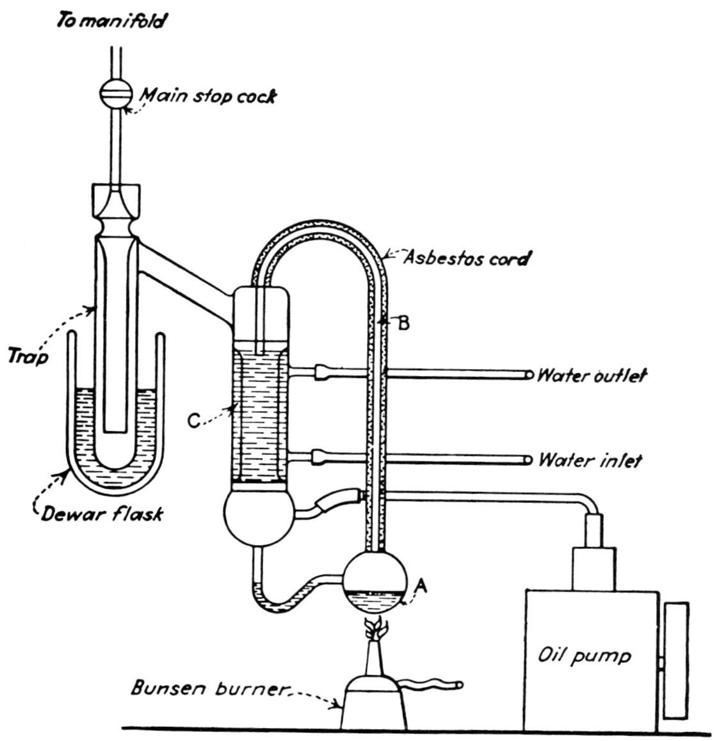

Figure 8.9. Water-cooled mercury diffusion pump with a vapor trap.

and its ruggedness make it preferable to the mercury type for this kind of work.

Diffusion pumps are usually made of pyrex glass, since this glass can stand heating much better than can lead glass. When such a pyrex pump is connected to the lead-glass manifold, either a graded-glass seal or pyrex-to-lead glass ground joint must be used to connect the pyrex to the lead glass.

The heater used is usually of the electric type, either internal or external, since this is convenient and compartively safe compared with a gas flame. But gas can be used just as well, provided that too intense a flame is not used. When mercury is used, the heat should be just enough to make the mercury bubble gently as it boils, while in the oil-condensation pump no visible boiling should occur. Figure 8.10 shows two special types of diffusion pumps.

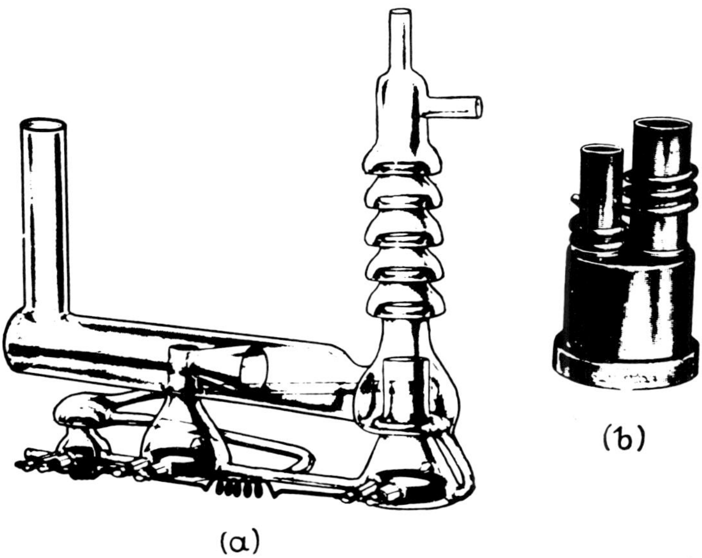

Figure 8.10. (a) Three-stage glass diffusion pump for very high vacuum. (b) Water-cooled metal diffusion pump.

There are several other names for the diffusion type of pump such as aspirator pump and condensation pump, but the pump is the same regardless of which name is used.

Vacuum Gauges. Besides the manifold and pumps already described, the vacuum system must have one or more vacuum gauges for measuring the degree of vacuum in the tube. All vacuum gauges used for luminous tube work operate on the barometer principle; that is, they are really pressure-measuring devices. The simplest type of vacuum gauge resembles the mercury barometer, as shown in Figure 8.11. This type of gauge, known as the *straight gauge,* consists of a straight tube at least 30 inches long and filled with mercury. The tube is fastened in an upright position over the mercury reservoir and is sealed into the manifold, as shown. When the manifold is full of air, all the mercury is in the cup; but when pumping is started, the mercury column rises in the tube about 30 inches. The height of the column below the zero level is an indication of the degree of vacuum. The tube is calibrated with a scale marked off in millimeters, which is fastened beside the tube. When the pressure falls to ten millimeters, the column is only ten millimeters below the zero level. This gauge indicates only a rough vacuum and is used to measure gas pressure only.

When the neon or other rare gas is put in the tube, its pres-

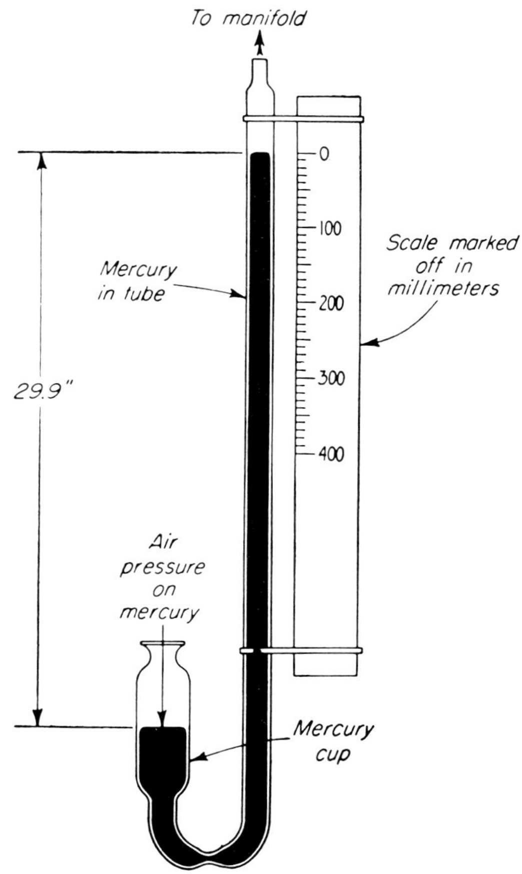

Figure 8.11. Straight type of mercury gauge.

sure may be measured with such a gauge. To secure a ten-millimeter pressure, the mercury column should fall ten millimeters; for a 15-millimeter fill, the column will fall 15 millimeters; and so on. For very low pressures of one millimeter or below, this type of gauge cannot be used, because the length of the mercury column is so long, and its changes in length are so small, that the pressure cannot be read accurately even with a magnifying glass. In other words, the straight gauge is a *millimeter gauge;* that is, it will read pressures from 1 to 760 millimeters but not below one millimeter. For reading microns, or thousandths of millimeters, a micron gauge is needed. Micron gauges are described in a later paragraph.

The U Gauge. The U type of vacuum gauge, shown in Figure 8.12, is a millimeter gauge also, but it is much smaller and

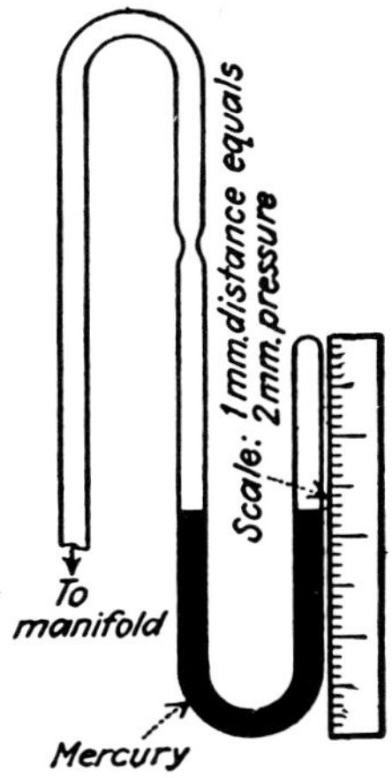

Figure 8.12. Standard type of mercury U gauge.

more convenient to install than the straight gauge. The U gauge is a U tube half-filled with mercury and sealed into the manifold system. One arm of the U tube is evacuated. When the system is pumped out, the pressure on the manifold side of the gauge is reduced, and the mercury in the two arms of the U tube tends to come to the same level. The difference in level between the two arms, in millimeters, is the pressure on the manifold side. The scale of such a gauge is placed beside the evacuated arm. The mercury in this arm will fall only ½ millimeter for every millimeter decrease in pressure in the manifold. This is true because the other arm rises ½ millimeter at the same time, making the difference in level 1 millimeter. Since the U-type gauge moves only half as far as the straight gauge, for the same decrease in pressure, it is less sensitive and harder to read, particularly for pressures in the region of 1 millimeter. But because of its simplicity and great usefulness as a vacuum indicator, almost all pumping systems are fitted with a U gauge. The U gauge will show very quickly just what the pressure is inside the manifold, and it will show also how quickly the pumps are working and if there is a leak in the system.

The mercury used in the straight and U-type gauges does not interfere with the drawing of a good vacuum, because the mercury is cold, and as a result, practically no mercury vapor is formed. For this reason, no special trap is necessary, as is necessary in the case of the mercury diffusion pump. Tubing used in gauges should be free from water vapor when the gauge is installed.

The Butyl U Gauge. A type of U gauge which is very useful in reading pressures from about ½ to 30 millimeters uses butyl phthalate instead of mercury. Since butyl phthalate is about one-fourteenth as heavy as mercury, the pressure inside the manifold will support a column of it fourteen times as high as it will support mercury; that is, a change in pressure of one millimeter in the manifold will cause a change of 14 millimeters in the height of the butyl phthalate column.

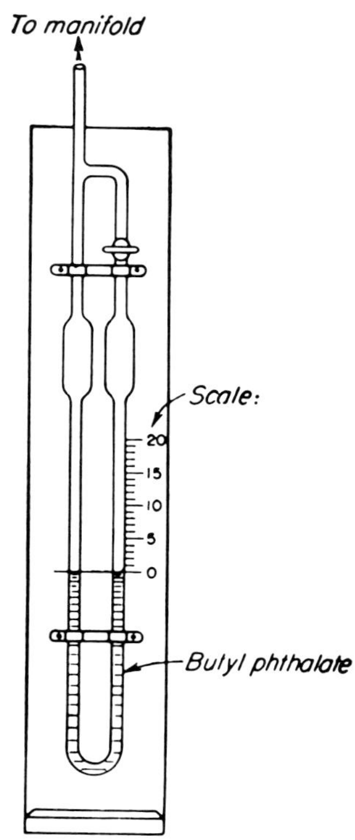

Figure 8.13. Butyl phthalate E-Z Read gas-pressure gauge.

As shown in Figure 8.13, this fact makes the scale of the gauge much wider and easier to read accurately. Filling to pressures as low as one millimeter and filling to higher pressures to an accuracy of 0.1 millimeter is entirely possible with such a gauge, while with mercury U gauges such accuracy cannot be attained. The butyl gauge is essentially a pressure gauge, useful for measuring the pressure of the rare gas, but not suitable for measuring vacuum pressures.

The butyl phthalate used here is the same substance as that used in the butyl diffusion pump, and it has the same advantages of low vapor pressure and chemical inactivity. The complete butyl gauge is about two feet long and fitted with a scale about eight inches in length.

The McLeod Gauge. The McLeod gauge is primarily a micron gauge; that is, it is designed to read very low pressures, as low as 0.1 micron. But the McLeod principle is also used in gauges designed for reading millimeters of pressure, and two gauges can be built into one, capable of both readings—high and low vacuums.

The McLeod gauge operates by compressing a sample of the gas from the manifold and comparing the pressure of this compressed sample with the manifold pressure itself. In Figure 8.14(*b*), the cup *C* holds a supply of pure clean mercury, while the rubber hose *H* connects the mercury reservoir with the gauge itself. The gauge is connected to the manifold at the point *A*. It is operated by lifting the mercury flask above the bottom of the gauge. The mercury from the flask flows through the hose and up the tube *T* to the point *B*. When it reaches this point, it divides into two paths. The left-hand passage (which contains gas at the manifold pressure) is cut off from the rest of the system by the mercury. As the mercury continues to rise in the tube *D*, the gas caught inside is compressed. At the same time, the mercury in the right-hand tube rises. When the mercury in this column has risen to the mark *M*, the level in the closed chamber *D* comes to rest opposite a number of the scale which shows the pressure of the gas inside the manifold.

The McLeod gauge does not read continuously; that is, each time a reading is desired, the mercury flask must be lifted, and the reading is taken after the mercury has risen to the proper height. The pressure reading indicates the pressure at the time when the mercury passed the point *B*, that is, when the sample in the chamber *D* was caught and the compression began. Reading pressures with this type of gauge is slow, therefore. In addition, the McLeod gauge cannot be used for measuring the pressure of the gas fill except by a cut-and-try method, which is slow and often very wasteful of the rare gas.

But for measuring very low pressure and for determining whether or not the pumping system can attain the required vacuum before filling the tube, a McLeod gauge can be used.

The range over which the gauge will indicate is determined by the dimensions of the chamber *D*. If a large sample of gas is caught and compressed into a very fine tube, the gauge can be made to read very low pressures. In the double gauge for reading both high and low pressures, shown in Figure 8.14(*b*), it will be seen that the micron part of the gauge has a very small bore compared with that of the millimeter section of the tube.

McLeod gauges must be very carefully made, and they must be calibrated; that is, the scale must be marked on them at the factory. It is very difficult for the glass blower to make his own McLeod gauge; the gauges are usually bought in complete form from a supply house.

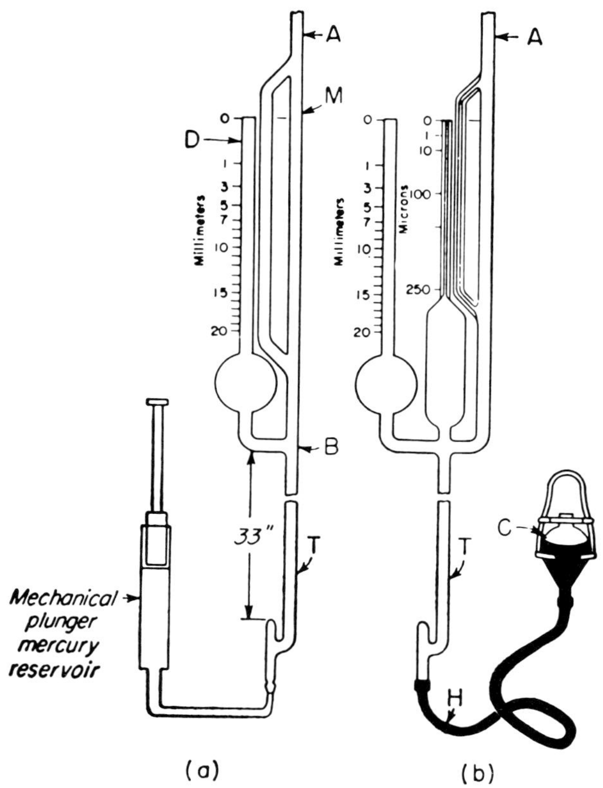

Figure 8.14. The McLeod plunger-operated gauge is shown at the left. The gauge at the right is the reservoir type and can be designed to indicate high vacuum in microns and gas pressure in millimeters. Either gauge is attached to the manifold by connecting to tubing at A.

Other Types of Vacuum Gauges. The McLeod gauge has certain disadvantages for a luminous tube plant. It is not instantaneous reading. The operator has either to push the plunger in the mercury pool or lift the flask of mercury to obtain the reading. Also, the McLeod gauge is made of delicate glasswork, and unusual jars or knocks may injure or break this glasswork.

During the last few years, more compact gauges have been designed and developed which have overcome some of the disadvantages of the McLeod gauge, and, more important, these can give almost instantaneous readings. The more commonly used gauges are the ionization vacuum gauge, the thermocouple vacuum gauge, the discharge gauge, and the Pirani type of gauge. These gauges have been built in various forms, but, in general, embody certain definite principles.

The ionization vacuum gauge uses a tube which consists of a filament, grid, and plate. The tube is connected to the manifold through a tubulation. The filament and the grid of the tubes are conventional, but the plate has a deposit of either silver or platinum on the inside wall of the glass. The lead is connected to this deposit and brought out with the leads of the grid and the filament.

There are two different circuit arrangements which can be employed. In one the grid around the cathode acts as internal collector and the outer coated plate acts as anode; in the other arrangement the roles are reversed, and the grid acts as the anode and the outer plate as the external collector. The pressure depends upon the degree of ionization produced in the low pressure gases left in the tube.

The underlying theory of the ionization gauge is that a stream of electrons from the cathode or filament ionizes the gas molecules to an extent depending upon the number of molecules encountered. When the pressure is low, or a high vacuum is reached, the rate of formation of positive ions will be proportional both to the number of electrons emitted from the filament and to the number of residual gas molecules left in the tube. In other words, at a high vacuum, the rate of ionization will be proportional to the degree of vacuum. The ionization gauge is not practical for neon or cold cathode manufacture because it has to be degassed when brought to atmospheric pressure.

Another gauge, similar to the ionization type but operating on a thermocouple principle, is known as the "thermocouple" gauge. A thermocouple tube is indicated in Figure 8.15 with a tubulation for sealing to the vacuum system. It contains a filament in the shape of an X having a small thermocouple welded to its mid-point. Four leads are extended from the envelope, two for the filament-heating current and two for the thermocouple. As the pressure is decreased, the energy removed from the hot thermocouple by conduction and convection is decreased; therefore its temperature and the thermoelectric voltage in-

Figure 8.15. Thermo-couple tube for vacuum gauge.

crease. The thermoelectric current output is measured by the "thermocouple" meter, which is a galvanometer. A thermocouple vacuum gauge is usually calibrated against a McLeod gauge. Once calibrated, the thermocouple gauge will remain accurate as long as the original arrangement is not appreciably changed. A thermocouple pressure gauge is practical to measure a vacuum from 2 to 200 microns.

The discharge gauge is a cold cathode, diode type of gauge and is non-burnout. One type of tube is enclosed in an oval-shaped metal shell. A permanent magnet produces a strong magnetic field along the minor diameter of the shell, establishing it as the cathode of the gauge. The anode consists of a rectangular loop of wire embracing the maximum portion of the magnetic field. As the pressure surrounding the gauge elements is reduced, that is, as the vacuum becomes higher, a discharge occurs in the gas. A direct-current millimeter, known as a "discharge" meter, reads the discharge current. These readings can be easily translated into accurate terms of pressure by calibrating against a standard, such as a McLeod gauge. The range of the discharge gauge is approximately from less than 0.1 to 20 microns. Figure 8.16 shows an R.C.A. vacuum gauge which incorporates in one unit both a thermocouple and discharge gauge for fully covering vacuums down to less than 0.1 micron. Figure 8.17 shows the calibration of both the thermocouple pressure gauge and discharge pressure gauge in "pressure in microns of mercury" against readings of the meters.

A gauge that is used more often than any other is the Pirani

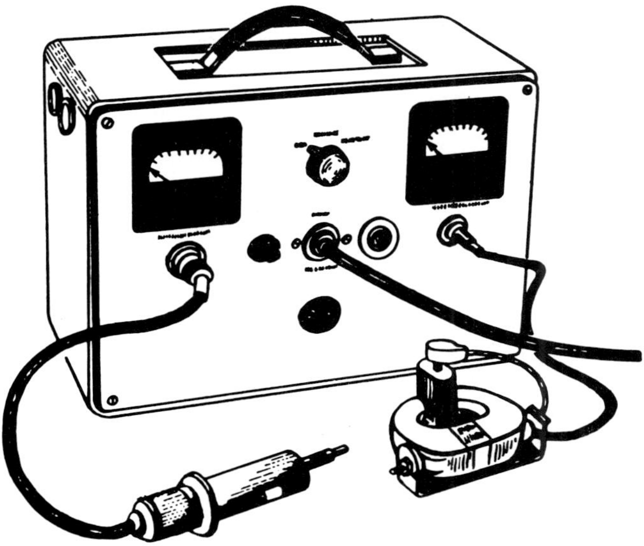

Figure 8.16. R.C.A. thermocouple gauge and discharge gauge incorporated in one unit.

type. Many modifications have been made in the original Pirani gauge. Some Pirani gauges use a single tube which is attached to the vacuum system. Another type uses two tubes, one of which is connected to the vacuum system and the other of which is a matching tube against which the one on the system is

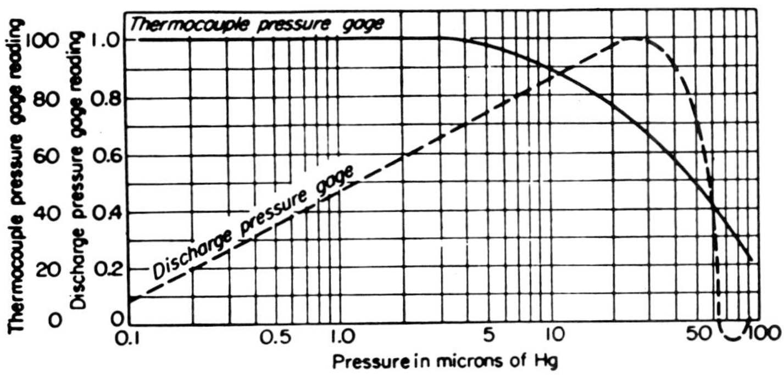

Figure 8.17. Calibration of the thermocouple gauge and discharge gauge shown in Figure 8.16. The thermocouple gauge operates in the range between 10 and 100 microns, and the discharge gauge from 10 microns down to 0.1 micron.

calibrated. The vacuum gauge indicated in Figure 8.18 is a modified form of the Pirani vacuum gauge. It operates on the principle that the heat conducted by a surrounding gas from

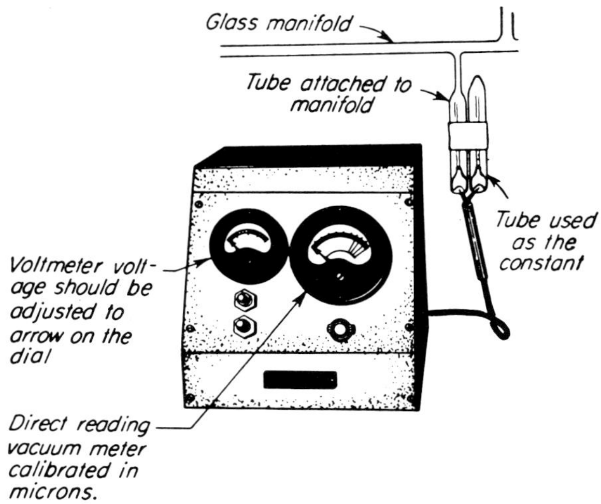

Figure 8.18. A modified type of Pirani vacuum gauge.

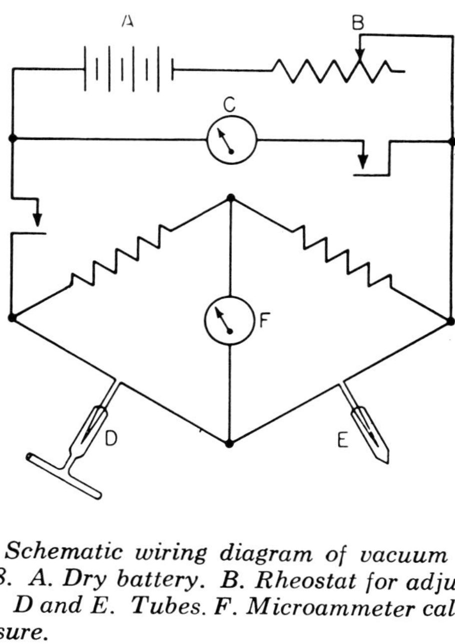

Figure 8.19. Schematic wiring diagram of vacuum gauge shown in Figure 8.18. A. Dry battery. B. Rheostat for adjusting voltage. C. Voltmeter. D and E. Tubes. F. Microammeter calibrated in microns of pressure.

an electrically heated wire, maintained at constant voltage, will cause a change in temperature of the wire and consequently will change its electrical resistance. This change in temperature and resistance of the wire will vary with the pressure of the surrounding gas and also with the kind of gas.

The use of the constantly heated wire as one leg in a bridge circuit which is balanced at atmospheric pressure makes it possible to detect a decrease in pressure because the bridge becomes unbalanced and the amount of the unbalancing current is measured in microamperes. The meter can be directly calibrated to read in units of high vacuum such as microns pressure. One micron represents 1/1000 millimeter of mercury. A circuit diagram of the gauge is indicated in Figure 8.19. To operate the modified Pirani gauge, it is necessary first to set the voltage accurately, as indicated by an arrow marked on the voltmeter. After the voltage is adjusted to the arrow, a button is depressed to connect the meter in the bridge circuit for a vacuum reading. A gauge of this type has many advantages in that the vacuum can be immediately read at any point. Another advantage is that, once the ultimate vacuum in the system has been reached, it is easy to determine whether there is any leakage in the system because of leaks or because of seepage through faulty stopcocks.

Trouble Shooting. If the system is not vacuumtight, it may be very difficult to locate the leak. A large leak may be detected by blowing into the system through the rubber blowing hose, but leaks large enough to be detected in this manner are usually visible to the eye. The smaller leaks, which prevent the system from attaining a good vacuum but which admit air very slowly, are much harder to find. This type of leak may sometimes be indicated by watching a U gauge and noticing if there is a gradual change in the two columns of mercury. A useful tool for detecting a leak is the high-tension *testing coil,* which has been referred to previously. This device is a transformer, mounted in a cylindrical case and equipped with a small knob (rheostat) at one end and a metallic electrode at the other. This tester, shown in Figure 8.20(*a*), is attached to the lighting circuit, and the knob at the end is turned until the electrode at the other end glows with a spark discharge.

Another type of tester for leaks which is used in the luminous tube field, shown in Figure 8.20(*b*), incorporates a high-frequency circuit in an insulated housing. This tester is furnished with a separate transformer. The advantage of the high-frequency test coil is that it can be used continuously without burnout. Also, because of the nature of the spark it produces, it will have less chance of puncturing the glass than the more commonly used unit.

Use of a Testing Coil. When the sparker is held against the glass tubing of the pumping system, the spark discharge will

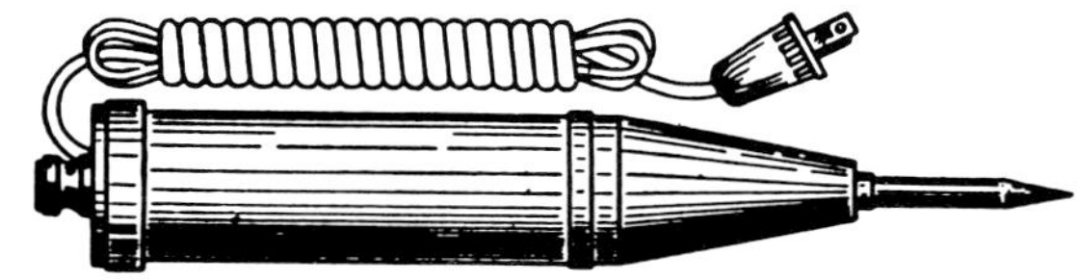

(a)

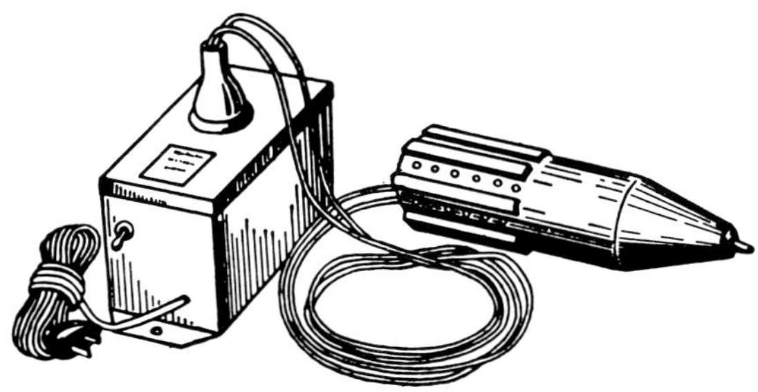

(b)

Figure 8.20. (a) Standard testing coil. (b) High-frequency tester.

jump to the glass and partially cover the outer wall of the glass. A poor vacuum will be indicated by a dark purple streak or red glow. If there is even a small pressure of gas inside the tube, a light blue glow will form inside the tube. As the pressure decreases, this glow will become thinner and will cling closer to the inside of the glass wall. When the pressure is below five microns, the glow inside the tube will disappear entirely.

The testing coil can thus be used as a rough indicator of the pressure inside the system, and it can indicate a very high vacuum by the absence of any glow inside the tube. But it is also highly useful for locating leaks. A pale red glow which gradually turns into a dark purple streak inside the tube indicates a leak. The spark from the test coil will enter the tubing through the leak, and in doing so it will become very white in color as it passes through the glass. The leak will be clearly indicated by the bright spot.

The test coil must be used cautiously, since the spark is capable of puncturing the glass, especially near a bend, splice, or thin part of the glass. The tip at the end of the coil should not be pointed, or else the spark will be very intense, and the tendency to puncture the glass will be much greater.

Sealing the Work to the Pumps. When the glass blowing has been completed as outlined in the preceding chapter, the electrodes are in place, and the tubulation is attached, the tube is ready for the pumps. The procedure of sealing the tube to the pumping systems is as follows: See that the stopcocks leading to the gas containers are closed and that the main stopcock leading to the pumps is closed. Then open the air-hose stopcock, admitting air to the manifold system. Break off a short section of the tubulation glass on the manifold.

Take the tube to be pumped, and lay it on the pumping bench so that the tubulations on the tube and on the manifold are close together. If they do not fit neatly, the tubulation on either the tube or the manifold may be heated and bent until a good fit is obtained. Then splice the two pieces of tubulation glass together, using the mouth blowtube to blow out the splice to the proper shape and thickness. When the splice is cool, blow on the air hose to test for any large leaks. If the tube and manifold appear to be airtight, close the stopcock at the air hose and open the stopcock leading to the pumps. This starts the pumping. Both mechanical and diffusion pumps should be operating at full capacity.

At the first instant the pump stopcock is opened, the sound of the mechanical pump will change to a deeper beat; but as the air is drawn out, the deeper sound will gradually disappear. As it does so, the mercury in the U-tube gauge will begin to come to an even level. When the mercury has come down the scale tube a short distance, close the stopcock leading to the pumps. If the system is airtight, the mercury will remain motionless at the level that it showed when the stopcock was closed. If there is a leak, the mercury will begin to climb in the scale tube. The leak should be located with the sparker. When it is found, the pump stopcock should be closed, the air hose opened, and the leak repaired.

When the system is shown to be completely airtight, the tube is ready for pumping. But before pumping is started, preparations for bombardment are necessary. These preparations, together with the detailed procedure of the bombardment process, are described in the following chapter.

Pumping Hints. Trouble with the pumping system may arise from a variety of sources. The following list of hints has been compiled to aid in running down the more common causes of trouble:

1. Keep stopcock surfaces greased with a thin coat of grease, and keep the grease out of the stopcock bore.
2. Keep grease, oil, and water out of the manifold system.
3. Keep the oil in the mechanical pump up to the proper level.
4. Use only the purest and cleanest mercury in the diffusion pump, if of the mercury type. If either organic liquid or silicone fluid, use pure solution.

5. Keep the pumps under vacuum at all times; that is, do not open the main-pump stopcock when the rubber-hose stopcock of the manifold is open to the air.
6. Keep the gas-container stopcocks closed at all times, except when actually filling. Never open the stopcock nearest the gas container until the piece of tubing between the two stopcocks is well pumped out.
7. Never turn on the heat under a mercury diffusion pump unless the water is circulating through the cooling system and unless the mechanical pump is running.
8. Keep flame away from the stopcocks. They will crack easily.
9. Keep the testing coil away from bends and splices except when testing for leaks.
10. Take down and clean the entire manifold once every four months.
11. Keep away from a mercury diffusion pump while it is operating. If it should break, the mercury vapor, which is highly poisonous, may get into the nose or mouth before it condenses.
12. Do not open the air-hose stopcock until all other stopcocks are closed. Do not accidentally open the pump stopcock while blowing in the hose; the suction may injure your lungs.
13. When using a flame on the system, never heat the glass to the melting point when the system is under vacuum except when sealing off a tube. The glass will suck in if melted.
14. Keep electric motor bearings well lubricated.
15. If a chain drive is used between motor and pump, keep fingers away from the chain.

Time Required for Pumping. The average time required for pumping lengths of tubing depends upon many factors that include:

1. The length and diameter of the tubing to be pumped.
2. The length and diameter of the piece of tubulation glass.
3. The length and diameter of the manifold glass leading to the pumps.
4. The diameter of the bore in the main stopcock.
5. The kind of pumping system used, either mechanical only or a combination of mechanical and diffusion pump.
6. The efficiency of the pumping system.

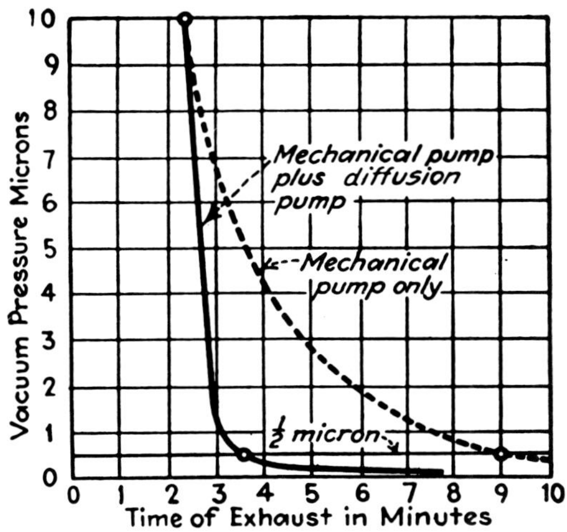

Figure 8.21. Comparison of pumping speeds between a mechanical pump and the same pump in series with a diffusion pump.

For an average diameter and length of tubing, Figure 8.21 indicates the average time required for pumping out this tubing with or without a diffusion pump. From this graph, it can be seen how valuable a diffusion pump is in speeding up production. The actual conditions in a particular plant may not require the time shown; in this case, a table similar to it can be made up, showing the actual time required to complete a tube of a given size.

CHAPTER IX

Bombarding

The Importance of Proper Bombarding. The commonest single cause of early failure of luminous tubing is improper bombarding or heat treatment. Nevertheless, bombarding is not a difficult operation; if the proper equipment is available, it can be made almost foolproof. It cannot be too strongly emphasized that a little extra time and care taken in properly bombarding a tube will pay big dividends in longer tube life, even though the difference may not be visible when the tube leaves the shop.

The immediate purpose of bombarding is very simple: It is to heat all parts (glass and metal) of the tube to a high temperature so that the absorbed gases and other impurities will be thrown off. Trouble arises when all parts of the tube are not brought up to this required degree of heat. When parts of the tube remain below the critical temperature, the gases and impurities in those parts are not eliminated, and in addition, these cooler parts will collect the impurities which have been thrown off from hotter parts of the tube. When the tube has been pumped, filled, and sealed, these impurities may lie dormant for many hours of operation. The tube may appear to be perfect when it is installed. But if the impurities are in the tube, they will surely be released, as soon as the operating temperature of the tube rises. The tube then loses color and must be repumped.

With the introduction of fluorescent tubing, proper bombarding became even more essential because of the critical nature of the fluorescent powder. The fluorescent coating is very sensitive to impurities left in the tube. Improper heat treatment may cause a decrease in illumination and in many cases darkening of the tubing. In signs, this will result in a loss of visibility, and in interior lighting it will cause the loss of illumination.

Bombarding Equipment. Success in bombarding is allied to success in glass blowing. Both depend upon the skill and judgment of the operator. Neither can be carried out by even the most skillful operator unless he has the proper equipment. Bombarding equipment is relatively simple, but it must be of sturdy design, and it must be suited to the job.

The commonest bombarding method, used almost exclusively in the luminous tube trade, makes use of a bombarding transformer which provides a high voltage and higher current than that of the ordinary tubing transformer. This transformer is connected to the tube while it is on the pumps and is used to produce a heavy glow discharge in the tube. It is thus the center of the bombarding setup. The power should be supplied preferably from the 220-volt power circuit. The transformer is supplied with an off-on switch and with some means of regulating the voltage it supplies. Lead wires for connecting the secondary terminals to the tube must be provided, and the entire installation must be insulated for the protection of the operator.

The Bombarding Transformer. The bombarding transformer is in many respects similar to the transformer used in lighting the tube (see pages 55 and 63). That is, the bombarding transformer has a definitely limited secondary current, even on short circuit. But the short-circuit current on the bombarding transformer is often ten times as high as the short-circuit current of the usual tube transformer. Tube-lighting transformers are usually made to deliver 120 milliamperes or less on short circuit; the bombarding transformer may deliver currents as high as 2000 milliamperes on short circuit.

The open-circuit secondary voltage of the bombarding transformer is usually 10,000 to 25,000 volts. The transformer, while similar in operation to the neon sign type, is much larger, since it must apply much more current. Furthermore, the bombarding transformer is a dangerous piece of apparatus, much more so than the tube transformer, because it has a higher secondary-current capacity.

The extra current is used for heating purposes when the tube is being bombarded. To heat the tube properly it is necessary to run through it many times the normal operating current, and for this reason the bombarding transformer must be able to supply a heavy secondary current. If the operator comes into contact with the terminals of the bombarding transformer or one terminal and ground, the full secondary current may pass through him; *and this current is definitely dangerous to life.* In order to guard against this possibility, thorough insulation of the entire secondary circuit is necessary.

The bombarding transformer is rated in kilovolt-amperes (abbreviation, kva). The rating is found by multiplying the primary voltage by the primary current and dividing by 1000. For proper bombarding, a transformer rated at least 7.5 kilovolt-amperes is required. If a smaller transformer is used, the number of feet of tubing which can be handled at the proper bombarding temperature is limited. Under such conditions, even average lengths of tubing will not come up to the proper temperature, and the bombarding is not fully effective.

Since bombarding transformers are relatively high-power devices, it is not economical to use them on 110-volt circuits. Usually, they are designed for 220 volts on the primary winding, since this reduces the primary current to one-half and the line losses to one-fourth. In almost every business or every industrial district, 220-volt circuits are available, and there is no reason why 220 volts should not be used for this purpose in almost every luminous tubing plant. In the rare case where only 110 volts can be obtained, a bombarding transformer can be supplied for 110-volt operation.

Bombarding Control. In order to control the bombarding current so the proper current will flow regardless of the length of the tube and of the pressure of the air in the tube, some sort of voltage control is necessary. There are three devices which can be used for this purpose.

The first type of control is the tapped reactor, or autotransformer. This device is a choke coil with an iron core. The coil is tapped at regular intervals, and these taps are connected to a set of switch points, as shown in Figure 9.1(*a*). The lever which

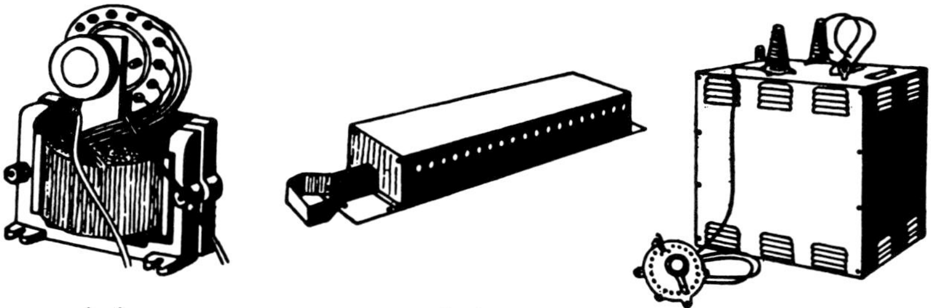

Figure 9.1. Bombarder regulators. (a) Rotary switch with taps on an autotransformer. (b) Reactor with a movable iron core for regulating current to bombarder. (c) Complete bombarder with separate rheostat for controlling direct current through reactor.

makes contact with these taps is used to control the part of the coil in shunt with the primary of the transformer and hence to control the power fed to it. This type of device wastes little power and is used to the exclusion of the simpler resistance rheostat, at least in the more modern plant installations. The switch handle of the bombarding control should be mounted conveniently so that the operator may use it without leaving the pumps.

The second type of control is a choke coil in series with the primary of the transformer. It is provided with a movable iron core. Very smooth control can be obtained by moving the iron core in and out of the choke coil [see Figure 9.1(*b*)].

The third type of control is known as the "d-c reactor"

method. In this method, shown in Figure 9.1(*c*), a series reactance is used as in the movable-core reactor. But the reactor in this case has an extra winding through which is passed direct current, the direct current being obtained from a vacuum tube rectifier not unlike the rectifier tubes used in radio receivers. The direct current changes the magnetization of the core and thus affects the passage of the alternating current through the other winding on the core. A small direct current is capable of controlling a very heavy alternating current in this manner. Consequently, the d-c rectifier and its controlling transformer may be of moderate size and hence convenient to adjust. The d-c reactor method of control has gained great popularity in recent years, especially where heavy bombardment schedules call for quick and accurate control.

Switches for Bombarding Work. The on-off switch is connected in series with the primary circuit of the bombarding transformer. The type of switch used varies widely, from a simple exposed knife switch or a knife switch enclosed in a metal box to a completely enclosed magnetic switch, with a remote-control push button. It is recommended that the bombarding switch be carefully installed. An open knife switch is dangerous for two reasons: First, it exposes the 220-volt line, which can give a painful shock, or cause a short circuit in the line. Also, it must be pulled open to turn the transformer off. An enclosed knife switch has its metal box grounded and is thus much safer. But the operator, leaning against a grounded box, may accidentally touch the high voltage terminal at the same time and receive a bad shock. A more desirable type of switch is the enclosed magnetic one shown in Figure 9.2 (*a*). This switch turns the transformer on when the button is pushed; when the button is released, the transformer goes off. If the operator accidentally comes in contact with the high voltage wiring, the transformer will be turned off as soon as he releases the button. For "Flashing," that is, turning the transformer on and off rapidly, this switch cannot be equaled. It is completely enclosed and safe.

Often a foot switch is used, since this leaves the hands free. But free hands can get into trouble while bombarding is going on, and it is usually wisest to use a hand switch.

The most satisfactory switch for bombardment control is the push-button controlled magnetic switch. The push button is connected to the magnet coil of the main switch and may be either 110 volts or 220 volts. The main switch is connected to the 220-volt circuit.

High Voltage Connections. Between the high voltage terminals of the transformer and the electrodes of the tube there must be two high voltage connecting wires. Since these wires are the part of the secondary circuit with which the operator may come in contact, they must be thoroughly insulated. The wire itself should be 15,000-volt sign cable, the same as is used in the sign

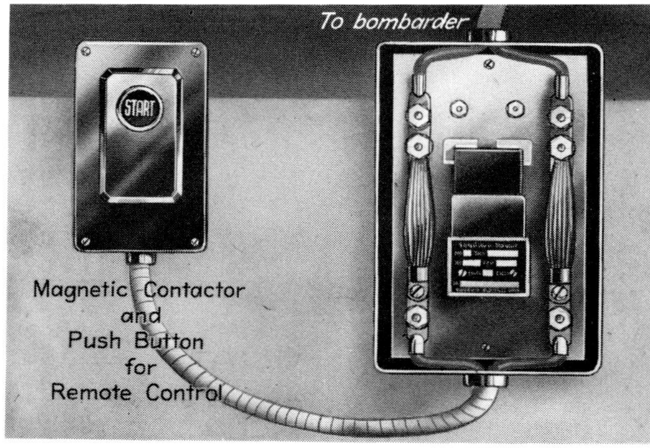

Figure 9.2(a). Magnetic switch operated by a push-button for bombarder operation.

itself. The clips which connect the cable with the lead-in wires of the tube should be of the spring-jaw type (commonly called *battery clips),* and they should be covered with rubber hoods, so that they can be attached without touching the metal of the clip. For convenience in moving these clips, two bare copper wires are often mounted on the ceiling of the shop, with rollers riding on them. Each connecting cable is attached to one of these rollers and thus can easily be pulled around to the proper part of the bench for connecting to the tube.

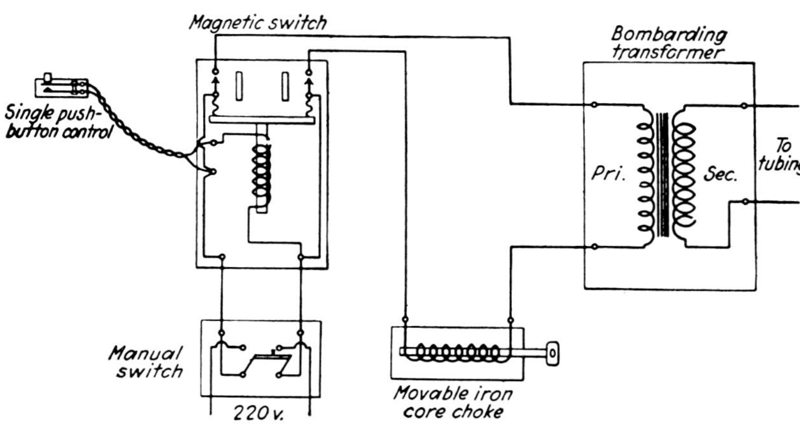

Figure 9.2(b). Wiring diagram for magnetic switch with push-button control.

Bombarding Layout. A convenient layout for the bombarding equipment is shown in Figure 9.3. The transformer itself with the magnetic switch is mounted to one side, out of the way, and provided with heavy leads which are covered with glass tubing. These leads run directly to the ceiling where they connect with two insulated copper bars. The connecting cables hang from pulleys which ride on the bare copper.

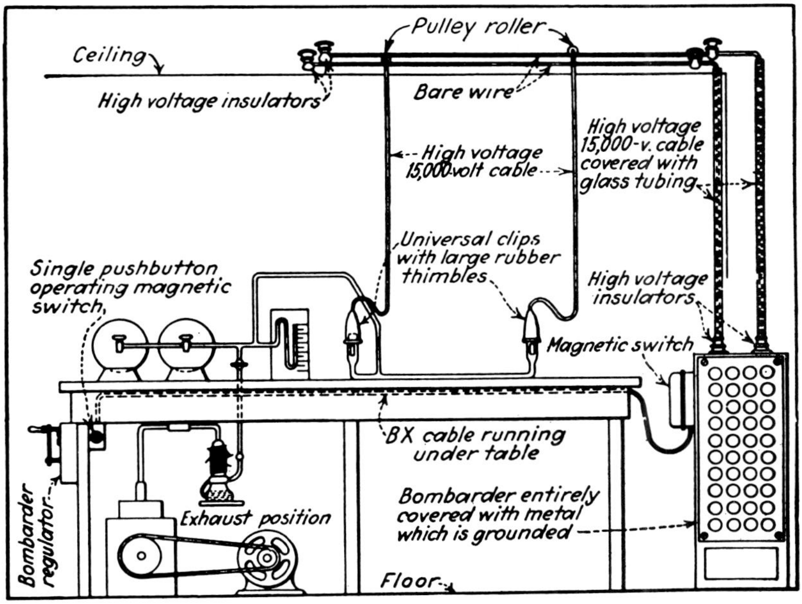

Figure 9.3. A typical bombarder layout. The high-voltage leads are connected to rollers which run on the bare wire fastened to insulators that are secured to the ceiling.

The bombarder regulator and push-button operating switch are mounted close together at the other end of the pumping bench, far enough away so that the operator is out of reach of the high voltage while the bombarder is turned on. These safety precautions should be rigidly observed. Men have been badly shocked from bombarding transformers; this would never have happened if the installations had been safeguarded and if the equipment had been used with care.

Bombarding Procedure, Simple Case. In the simple case of direct bombarding, the following is the correct procedure: The tube is sealed to the pumps, and the entire system is tested for leaks, in the manner outlined in the previous chapter. When the system has been made vacuum tight, air is admitted to the manifold by opening the air-hose stopcock. This stopcock is then closed.

The high voltage wires from the bombarder are connected to the electrodes of the tube. Next, a temperature indicator described later is placed against the side of the glass tube so that it makes good contact. The main-pump stopcock is then opened (with both mechanical and diffusion pumps operating). The bombarder control is set at the minimum current. As the pressure goes down, as indicated by the U gauge, the bombarder should be flashed, by giving the push button a momentary push, until a flash of light occurs in the tube. When this happens, the pump stopcock is closed immediately and the bombarder circuit kept closed. The bombarder current is increased slowly by regulating the choke. If the tube glows steadily, well and good; if not, the pump stopcock must be opened for a fraction of a second until a steady glow is obtained, keeping the stopcock closed.

The bombarder should be kept on and the current increased until the tube gets thoroughly hot. This is indicated by the temperature indicator placed upon the glass tubing. The electrodes must be watched carefully and should not be allowed to get too hot before the glass has been thoroughly processed. For proper bombarding, the glass should get hot first and then the electrodes.

When the proper temperature of the glass has been reached, the pump stopcock is opened and bombarding continued with the stopcock open. The tube will glow as before, but the electrodes will get much hotter. While the pumping is going on, they should be raised to a high red heat but not overheated. It is especially important to get the electrodes to a high heat, not only to eliminate the impurities from the metal itself, but also to break down the electrode coatings. *This is very important.* Electrode coatings which have not been broken down may cause a great deal of trouble by releasing impurities, causing contamination especially in fluorescent tubing.

Laying paper on the tubing for testing temperature by charring the paper is a rough method for determining the degree of heat. A better method is to use a meter that will accurately indicate the true temperature of the glass while bombarding. Figure 9.4 shows a temperature indicator that consists of a millivoltmeter calibrated in degrees centigrade, connected to a pair of leads ending in a thermocouple that is clipped onto the glass tubing. The two wires are iron and nickel-iron alloy. A change in temperature of the glass produces a small change in voltage at the thermocouple and corresponding reading on the meter. For the average neon sign or cold cathode tubing, a temperature of 200 to 225°C is required for good heating; the maximum amount of gases are released and eliminated at this temperature range that may effect the operation of the tubing later on.

When the bombarding is completed and after pumping has taken place for a short period, some pumpers test the vacuum in the tube by momentarily closing the bombarder circuit when the pressure is very low. If a discharge occurs, it is taken as an indication that there are still impurities in the tube. If no discharge occurs, it is taken as an indication that the vacuum has reached a fairly high degree. It is best to test for a vacuum with some approved gauge rather than with this method.

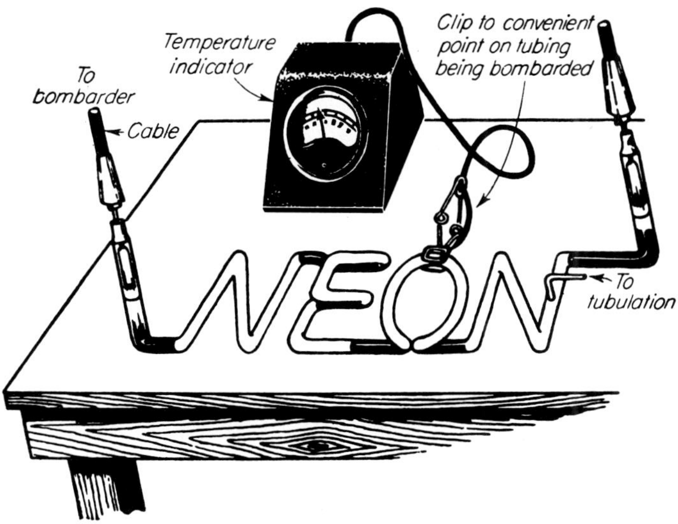

Figure 9.4. A temperature indicator connected to glass during the bombarding process.

When a high voltage test is made, do not keep the bombarder on if there is a low pressure in the tube. Otherwise, the electrodes may be seriously injured or the discharge may attack the electrode wires in the rear of the shell. Usually one good bombarding treatment is all that is necessary with a fast pumping system, especially for standard lengths and diameters of luminous tube units. In many cases where there are some unusual lengths of tubes or where tubing of larger diameter and length is bombarded for interior lighting use, a second bombarding treatment is recommended. Any of the rare gases such as helium, neon, and argon can be used for the second bombardment with special emphasis placed on thoroughly heating and degasifying the electrodes.

Some manufacturers recommended what is known as open-stopcock pumping and bombarding. This means that the bombarding is carried on with the main stopcock open all the time.

Unless the electrodes are heavily coated, this method is not too satisfactory because it prevents proper heating of the glass. The heating of the glass depends largely upon the amount of impurities left in the tube and can be accomplished only with closed-stopcock bombarding. With open-stopcock bombarding, the pumps will pull out gas faster than it is generated from impurities released from either the glass or the electrodes. No matter what method of bombarding is used, after the operator is satisfied that the proper heat treatment has been applied, the bombarder wires can be disconnected. The tube is then ready for filling.

Bombarding Cold Cathode Fluorescent Tubing. Although the above procedure for bombarding can be followed with almost all types of tubing, it is necessary to take more precautions when bombarding cold cathode tubing than ordinary sign tubing. Light output is the paramount quality required of cold cathode tubing. This can be badly affected by improper bombarding and improper processing. It becomes important to take more care in processing such tubing so that the maximum light output will be obtained.

Sometimes when open fluorescent tubes have been lying around, they absorb a quantity of moisture. If a tube with traces of water vapor in it is bombarded, the fluorescent coating may be attacked. A good procedure when bombarding cold cathode tubing is first to bombard and heat the tubing to a warm temperature. This should take about ten seconds. Then allow the pump to pull out all the impurities, especially the water vapors released during the warming-up period. By doing this, the water vapor will not combine with the fluorescent powder when the final or intense heating is applied to the unit. After waiting a half minute for pulling out the water vapor, the tubing can be filled with either air or gas and the standard bombarding procedure applied, as indicated in previous paragraphs.

Bombarding Current. Every bombarder should have a milliammeter in the secondary circuit connected in series with one of the bombarder leads. This milliammeter will indicate the secondary current which is flowing through the tube during the bombarding treatment. For all practical purposes, the milliammeter can be one reading from zero to 1000 milliamperes. The bombarding current should start at approximately 200 milliamperes for small tubing up to 12 millimeters in diameter. It should go up to approximately 400 milliamperes when the intense heat is applied. For 12 to 15-millimeter tubing, the intense heating current should be approximately 500 milliamperes. For tubing 18 to 25 millimeters in diameter, the intense heating current should be 600 to 750 milliamperes.

Reasons for the Bombarding Procedure. The above bombarding procedure for the simple case is based on sound practice. If followed, it should give complete and satisfactory bombarding under all conditions. The tube should be bombarded at the highest possible pressure, that is, just after it begins to glow and before the pumps have had a chance to reduce the pressure below that point. The reason for bombarding at high pressure rests on the fact that the *glass* will be heated only if there is *plenty of gas* inside the tube to carry the heat to the glass wall. In the high pressure glow discharge, most of the heat goes into the gas itself, and the gas transfers it directly to the walls of the tube. If the gas were at lower pressure, the electrodes would not become hot. To heat the glass, bombard at the highest pressure at which a steady glow will occur. To bombard the electrodes, a somewhat lower pressure should be used, but not too low, or the electrodes will be heated excessively.

Care must be taken not to leave the stopcock closed too long when bombarding with the stopcock closed. If too much gas is generated by the heat of bombardment and not removed, black rings may appear throughout the glass tubing. Discretion must be used, therefore, in momentarily opening the stopcock at appropriate intervals to allow the impure gas to escape.

When purchasing a bombarder transformer for the plant, care should be taken to obtain a unit large enough not only for the present output of the plant but also with allowance for future expansion. In the small plant, a bombarder rating of 7.5 kilovolt-amperes is large enough. But the growing popularity of large-diameter tubing, from 18 to 25 millimeters in diameter, may bring the need for larger bombarding units. Larger bombarding transformers are available in ratings from 12.5 to 22.5 kilovolt-amperes.

Bombarding Gas. As explained previously, it is wise in many instances to bombard a second time by using a rare gas. Other gases are available, such as hydrogen or a mixture of hydrogen and helium, which are furnished in containers like those for rare gas. They may be used for bombarding purposes, especially for bombarding dirty tubes which have been used on signs for a considerable length of time. These gases should be used with discretion.

In many cases, the operator will insert a quantity of ordinary illuminating gas for bombarding a particularly dirty mercury tube. By warming the tubing in the usual way, especially when mercury vapor is present in the tube, and then inserting a small quantity of illuminating gas before heating to a high temperature, the black or dirty sections of the tubes can be eliminated. A few fillings may be necessary to clean up a particularly dirty tube. Extreme caution must be employed in the use of illuminating gas for bombarding.

Helium Electrodes. With certain types of electrodes specially developed for helium gas, it is very important not to exceed a certain current during bombardment. These electrodes have been specially processed and are usually kept moistureproof. If too high a current is used, part of the electrode coating may be knocked off and deposited on the glass, causing cracks. The manufacturer's instructions which accompany such electrodes usually give directions for properly bombarding these units.

External Bombarding. The method of internal bombarding by heavy current-glow discharge, described above, is in almost universal use in luminous sign work. But it is not the only method. For some special work, the process known as external or high frequency bombarding is used. This type of bombarding is suited for heating the electrodes only; when this method is used, the glass tube either must be baked in an oven at 850°F to eliminate the impurities from the glass or must first be bombarded by the internal bombarding method. The metal in the electrodes is then heated by a process of *electrical induction,* similar to that used in the high frequency electric furnace. A heavy coil of copper tubing is connected to a generator of high frequency electric current, similar to that produced in radio transmission. The copper coil is held over the outside of the glass so that it surrounds the electrode. When the current is turned on, a high frequency current is induced in the electrode metal. As a result, the metal of the electrode becomes very hot; in fact, a white heat can be produced if the current is high and kept on too long.

Bombarding Tests. There are a few tests which may be applied to the tube before it is filled to determine whether the bombarding is sufficient. They all involve measuring the pressure in the tube after it has been thoroughly pumped down. A high-vacuum gauge may be used to measure the vacuum, if one is attached to the system. If not, the spark-test coil may be used for a rough test. If the pressure has not been reduced to a low point, it may be that there is a slight leak or that the pumps are not capable of producing the required vacuum. But if the pumps are in good working order, high pressure will usually indicate improper or incomplete bombarding. Filling the tube with air or a rare gas and starting again from the beginning is often the only remedy.

The degree of heat necessary to liberate the absorbed impurities in the tube and electrodes depends upon the type of material and the kinds of impurities to which it has been exposed. The glass itself contains two different kinds of impurities, the surface particles and the absorbed gases (see Figure 9.5). The surface particles can be either converted into gas or entirely degasified by the heat of ordinary bombardment. The gas which is within the structure of the glass itself requires very high heat for elimination. Since these gases do not affect the operation of the tube, normally, they are not removed.

Usually the proper heating treatment as applied to the glass itself is satisfactory to eliminate all the impurities that may be within a fluorescent coating applied to the inside of a glass tube.

Caution Required. If the bombarder is used carelessly, it will cause a great many tubes to be ruined. This is particularly true when the tube is being bombarded at low pressure after the glass walls and electrodes have been thoroughly heated. If the bombarder is kept on too long when the pressure is low, the electrodes may become very hot and may even melt. More often, however, the discharge reaches the lead-in wires behind the electrode and either melts or weakens them or else heats them to such a degree that the seal of the tube is cracked. Also, in many cases, the

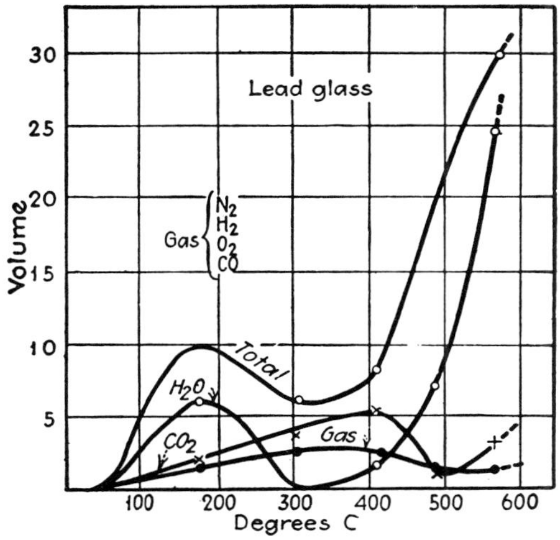

Figure 9.5. Curves showing gases released from lead glass heated to various degrees of temperature. The following symbols are used: H_2O water vapor; CO_2 carbon dioxide; "gas" N_2 nitrogen; O_2 oxygen; CO carbon monoxide. Note that above 400°C the released gas comes from within the glass structure. (Curve due to Sherwood, Westinghouse Electric and Manufacturing Company.)

electrode glass jacket may be darkened by a metal deposit coming off the shell or lead-in wire. The bombarder should be kept on only for a second or two at a time except when the high pressure is still in the tube. Good practice in bombarding also requires safety precautions for the operator. A heavy rubber mat placed on the floor in front of the bombarding equipment will reduce the hazard of accidental shocks.

CHAPTER X

Filling, Testing, Aging

Filling the Tube. When the tube has been bombarded and pumped to a pressure of five microns or less, it is ready for filling with the rare gas. Some mention has already been made of the flasks which contain this gas, how they are attached to the manifold system, and how the gas is allowed to flow into the tube until the proper filling pressure is reached. This procedure is simple enough, since it involves only opening one or two stopcocks and keeping an eye on the pressure gauge at the same time. But if the table is to be bright and of the proper color, great care must be taken to protect the rare gas inside the flask, so that it does not become mixed with air or other impurities. If the stopcocks are opened at the wrong time, for example, when the manifold is not pumped out, the impure gas from the manifold may mix with the rare gas. When this has happened, the gas in the container is ruined, and there is no way of repairing the damage except by installing a new flask. It is well to understand the containers, therefore, before using them to fill a tube.

Rare-gas Containers. The standard gas container shown in Figure 10.1 contains somewhat more than one liter (1.05 to 1.10 liters) of rare gas at atmospheric pressure (760 millimeters). The container has two seals: one on the end of the round globe and one in the "neck." The seal in the neck is broken after the container is sealed to the system and after all air and impure gas between the seal and the stopcock have been pumped away.

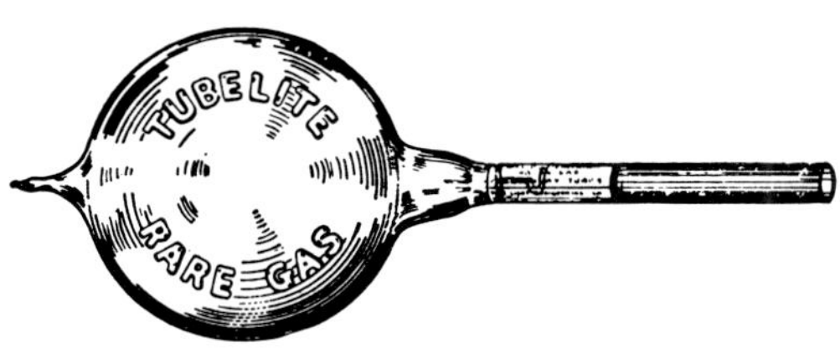

Figure 10.1. Ball-type rare-gas container. The fine tubulation in the neck is broken after the container is sealed to the manifold and the neck portion of the container has been thoroughly evacuated.

When the bottle is first bought from the manufacturer, it is common practice to test it to see whether it is full and whether the gas is pure. Although every precaution is taken by the manufacturer to ensure a full container of pure gas, the possibility of leaks developing in the flask during the shipment makes it is advisable to test the container before sealing it into the manifold.

For years it has been standard practice to test gas bottles with the spark-coil tester. When the test coil is held against a neon container, the bottle should show well-defined streaks, or streamers, of reddish color. If these streaks appear, the bottle is accepted as in good condition; if not, it is rejected. It has been shown that this test is by no means a sure indication of the purity or quantity of the contents. In one test, two separate liters of neon were checked by the spark-coil method. The first bottle showed streamers of red; the second, merely a red glow. When both bottles were opened and checked in a laboratory, it was found that they contained identical amounts of pure gas. One of the reasons why the spark-coil test is not always accurate is the fact that the sparker has an adjusting rheostat which permits a different intensity and frequency of spark. Even small changes in intensity or frequency makes a great difference in the character of the glow in the bottle.

It has been found that the spark test can be used only for testing color. A pure red glow of any sort, whether streamers are present or not, indicates that the neon in the container is pure. If a bluish color appears, then it is certain that the neon has become contaminated with air or other impure gas. It has also been found that it is not wise to test a bottle with the sparker more than two or three times. Under the influence of the spark, gases in the glass wall of the container may become free and contaminate the neon gas. Care should also be taken to avoid puncturing the glass of the container with the spark coil.

When testing a gas container with the test coil, it should never be applied to the neck of the container, or the internal seal may be punctured. The puncture may be so small that it is invisible, and the leak may not show any visible effects for several hours.

Installing the Flasks. The following method should be used in installing the rare-gas bottle on the pumping system: Lay the gas container on the workbench, with the bulb in a box or other support so that the neck of the container is parallel to and several inches above the surface of the bench, as shown in Figure 10.2. Procure a glass stopcock of two to four-millimeter bore, preferably of the long-barrel type. It is essential that the stopcock be of the vacuum type, thoroughly ground, so that both the cock and seat fit together without any microscopic leaks. The stopcock should be greased according to the instructions given on page 139, and particular care should be taken

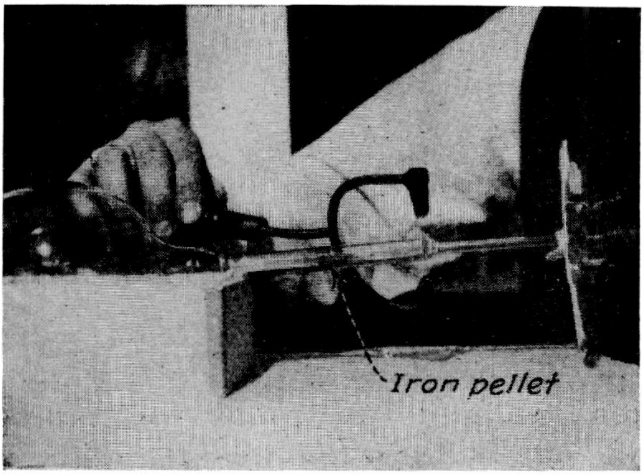

Figure 10.2. Sealing a stopcock to the rare-gas container. The asbestos shield to the right of the torch is used to protect the stopcock from the heat of the flame. Care should be taken to prevent the iron pellet (see arrow) from breaking the seal of the container during the sealing operation.

with the stopcock which is sealed directly to the rare-gas flask. Next prepare a small solid piece of glass or a piece of soft iron. If iron is used, heat it in a hot flame to remove any surface impurities, and then allow it to cool. Slip the glass or iron piece into the neck of the gas flask, then seal the stopcock to the bottle. Be very careful not to break the delicate seal inside the neck. Seal the other end of the stopcock either to the vacuum system or to the other stopcock. Start the pumps, open the stopcock, and pump out the entire system up to the inside seal of the container to as good a vacuum as the pump will draw, that is five microns or less.

Close the stopcock and cut the container and stopcock off the system. Then lift the container and allow the piece of iron or glass to fall on the seal. Shake the container until the glass seal is broken. The gas will then flow out as far as the stopcock. The entire container with stopcock is then sealed back to the vacuum system and is ready for use.

If it is not convenient to remove the container from the system in order to break the seal, the iron piece may be lifted by a small hand permanent magnet or by an electric one, and allowed to drop on the seal without moving the container from its position in the system. Only an iron or steel piece can be used with the magnet for this purpose.

If two stopcocks are used on each container, as is the best practice, there should be about four to six inches of tubing between them. They may be sealed in place before the pumping;

or a single stopcock may be sealed on as outlined above and the extra stopcock sealed on afterward.

Purity of Gas. The important thing to remember in opening and sealing on the gas bottles is to keep the gas inside the container pure. The seal should not be opened until the space on the manifold side of the seal is under a good vacuum, and the stopcock should never be opened after the seal is broken, unless the tubing beyond the stopcock is well pumped out. If air is admitted to the container under any condition, it will mix with the gas and ruin it for luminous tube purposes. The gas will flow out if the pressure is higher inside the container than out. Air will flow in if the pressure in the container is lower than the air pressure, as it is after the container has been in use for some time.

Since the performance of the ultimate luminous tube depends on the purity of the gas, every precaution should be taken to protect it from contamination.

In choosing the rare gas, reference should be made to Tables 3.10 and 6.1, which give the gas or mixture for various colors, together with the recommended pressures.

System Using a Metal Gas Cylinder. Many shops that manufacture neon signs in large quantities may find the constant installation of single glass containers a hindrance in rapid production of glass tubing. For these quantity producers, a system has been developed utilizing rare gas under pressure (approximately 250 pounds per square inch) in a metal cylinder. The metal cylinder is connected to a single liter glass container by copper tubing and glass connections, as shown in Figure 10.3. A metal cylinder may contain 15 liters of rare gas and can be made to hold more if required, depending on the size of the cylinder and the pressure used.

In a system of this type, it is essential to have all the connections vacuumtight, such as the nut and nipple connection to the metal cylinder and the metal-to-glass (de Khotinsky) joint. With the valve on the metal cylinder firmly closed, evacuate the entire system with all the stopcocks open. The mercury in each of the two columns of the manometers will rise approximately 29 to 30 inches. All glass stopcocks are then closed, and the system is tested for leaks. When the absence of leaks has been established, the first stopcock is then opened and the valve on the cylinder is opened carefully. The high pressure in the cylinder will force the rare gas into the glass container and cause the mercury columns to fall. A very small hole has been provided in the nipple of the cylinder valve to prevent a too rapid flow of gas into the glass container. When the mercury in each of the tubes has fallen to equal level with the mercury in the large pool, then the glass container has been filled to one liter at atmospheric pressure. The metal cylinder valve is tightly closed, and all the stopcocks are also closed. The second and third stopcocks can then be used for ladling in the gas to the tubing as usual. The mercury manometer to the right will indicate when the filling procedure must be repeated.

Filling the Tube. When the bombarding procedure, as outlined in the last chapter, has been completed, the tube must be thoroughly pumped out before filling. This pumping-out process should not take more than a few minutes, but at the end of that time the pressure should be below five microns. If a vacuum gauge is available on the system, it should be used to test the vacuum before filling. If a gauge is not used, the test coil can be used on the tube while it is being pumped. If any continuous glow appears inside the luminous tube, the vacuum is not low enough (see Table 10.1).

In some cases it has been the practice to compensate for a poor vacuum system by filling the tubes to a lower rare gas pressure than that recommended in Table 6.1. This practice will produce tubes that age up quickly, but whose life is short, often only one-tenth of what it might be if good pumps and a standard gas pressure were used. This point cannot be stressed too much. *Pump the tubes down until only a barely visible glow*

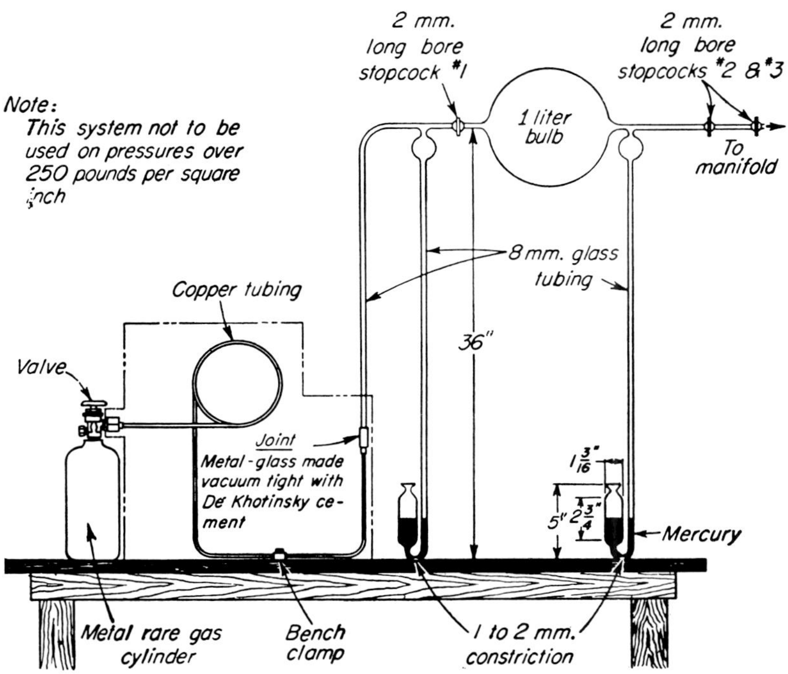

Figure 10.3. Typical system used when rare gas is furnished in a metal cylinder up to 250 pounds pressure.

Table 10-1. Degree of Vacuum by Color, Using a Spark-coil Tester

Degree of vacuum	Color	Coil held against: Glass	Coil held against: Electrode wire	Remarks
200 millimeters	Purple	..	✓	Color visible only at edge of electrode shell
150 millimeters	Purple	..	✓	Faint glow in tube
50 millimeters	Purple	..	✓	Fair glow in tube
15 millimeters	Blue-purple	..	✓	Dark glow
4 millimeters	Red-purple	..	✓	
2 millimeters	Lavender	..	✓	
1 millimeter	Light lavender	..	✓	
250 microns	Dark blue	✓		
100 microns	Light blue	✓		
10 microns	Very light blue	✓		
5 microns	Blue disappears	✓	..	No glow in tube; blue haze near glass wall

occurs when the test coil is applied. If the pumps will not reach this degree of vacuum, or if they require too long a time to reach it, there is something wrong with the system. It is false economy to use a pumping system in poor condition, because considerable aging may be required for standard pressure tubes, or short-lived tubes may be produced, or excessive darkening may occur in mercury-filled fluorescent tubes.

Measuring the Rare-gas Pressure. The filling pressure to be used in the tube may have been specified when the original layout was drawn up, according to the recommended practice. Often, however, the pressure is left to the discretion of the pump man, who must know how much pressure to use to produce the desired operating current and tube life.

When the tube has been pumped out to the required degree of high vacuum (five microns or less), the main-pump stopcock is closed. Be sure that the pump stopcock is closed before filling; otherwise the rare gas will be pumped out of the system in a very few seconds. The pressure gauge should read approximately zero millimeters, since the pressure is at five microns (only 0.005 millimeter). The stopcock of the bottle containing the gas used to fill the tube is then opened slowly. When the gas begins to flow into the tube, the pressure gauge will rise from zero, in succession, to 1, 3, 5, 8, 12 millimeters, indicating that these pressures have been reached inside the tube. When the desired pressure has been reached, the stopcock is turned off.

If too much gas has been admitted and the gauge reads high, the pump stopcock may be opened very slowly until the pump has drawn out the surplus gas and the pressure has been reduced to the desired value. This procedure is wasteful, of course, and should be avoided unless absolutely necessary.

The use of two stopcocks on each gas container, besides being an added precaution against leaks, helps to ensure filling to the proper pressure without going over the mark. When two stopcocks are used, the lower one (nearest the manifold) is opened, so that the space between the two is thoroughly pumped. The lower stopcock is then closed, and the pumps are turned off. The upper stopcock (nearest the gas bottle) is then opened, admitting a small portion of gas to the space between the two stopcocks. The upper stopcock is then closed and the lower one opened, thus admitting the gas to the tube. The first portion of gas may not be enough; if so, a second sample is taken by the same process. After a little practice, the pump man will find that he can "ladle" gas into the tube very accurately without wasting it.

A more elaborate setup for admitting gas to the tubes uses three stopcocks between the gas container and the tube. The stopcock next to the container is always kept open, while the other two are used for ladling the gas in the manner outlined in the preceding paragraph. The third stopcock is closed only when the other two are cleaned or regreased.

Testing the Tube on the Manifold. In many cases, it is desirable to test the tube while it is still on the manifold. This will show whether it is perfect or will require aging. If it is not up to standard, it may be repumped and refilled while it is still on the pump. When the pump man is thoroughly familiar with his pumps, such tests are rarely necessary, but occasionally they save a lot of trouble and effort, especially if the pump man suspects that all is not well with the tube.

One method sometimes used for testing the tube after rare-gas filling, while it is still on the pump, is to connect the bombarder to the tube and to flash it once or twice on a low bombarder current. This test is simple to make, but it is likely to spoil the tube if the bombarder is left on too long at a high current. It should be flashed on and off quickly.

Sealing Off the Tube. If the tube tests well, or if it is judged by the pump man to be satisfactory, it is sealed off the system as follows: The tubulation glass near the tube is heated uniformly around the entire diameter with the hand torch for a distance of about 1/4 inch from the tube. Care should be taken to heat the tubulation tube evenly and slowly until it begins to melt. When the glass is plastic and is being pulled in by the vacuum, quickly draw the tube away from the manifold, thus pulling out the melted glass and sealing the tube. When the glass has been pulled out about an inch, seal off the glass as close as possible to the tip of the tube. Do not leave sharp points on the tube or the manifold tubulation glass.

It is wise to practice tipping off seals on the vacuum system before attempting to seal a completed tube, since the process is a delicate one. If the glass is heated too suddenly, it may crack. If it is heated too much in one spot, that one spot will melt first and suck in, causing a messy seal or possibly a puncture in the glass. If too heavy a seal is made, strains are set up in the excess glass causing the tip to break off afterward, often after considerable time has elapsed.

Filling Mercury Tubes. The procedure which has been outlined in the foregoing pages for pumping, bombarding, and filling applies to all tubes, regardless of the color or gases which they contain. The mercury tube is filled with its carrier gas in the manner described, but, in addition, it has to be filled with a small amount of mercury itself. As stated on pages 13 and 14 the mercury tube is in a class by itself, because of the liquid mercury that it contains. Mercury is chemically active and will combine with the smallest amount of dirt or impurity present in the tube and form black spots, streaks, or rings inside the tube. In addition, the mercury vapor which provides the light of the tube is temperamental; it will not function well in cold weather. These special characteristics of the mercury tube make it necessary to follow special precautionary measures in filling the tube.

Many factors must be followed in order to produce a clean mercury tube. The first requirement for a good mercury tube is clean glasswork. Clear glass should be thoroughly washed and cleaned before it is used or coated with fluorescent coating. The second requirement is that the glass, particularly fluorescent glass, should not be allowed to accumulate any dust, dirt, or moisture. The third requirement is particularly thorough processing or bombardment with special emphasis placed on thoroughly breaking down the coating inside the electrode shell. This can be accomplished only by intense heating of the electrode to a high temperature. The fourth requirement is inserting absolutely clean mercury and keeping it pure and clean after it gets inside the tube. The fifth requirement is to insert the correct amount of mercury. The sixth requirement is filling the tube with the proper carrying gas to the proper pressure.

The first, second, third, and sixth requirements can be met with patience and careful work, but the fourth and fifth cannot be met without special mercury-inserting equipment.

Methods Used for Mercury Insertion. Two or three methods are commonly used for inserting mercury into tubes. Some manufacturers use a mercury reservoir on the pump system, fitted with a special stopcock. The hole in the plug is drilled only one-half of its diameter, leaving a small reservoir. When a tube is to be filled with mercury, the stopcock is turned 180 degrees, and a few drops are allowed to flow by gravity into the tube. This system has two disadvantages: the mercury in the reservoir cannot be kept absolutely pure and clean; and, in passing through the stopcock, it is almost certain to pick up stopcock grease and dirt. Furthermore, it is difficult to measure a small amount of mercury by this method; usually the tubes so made contain an excess of mercury, which, in fluorescent tubes, may be harmful.

Some tubing is made by simply pouring the proper amount of mercury into the tubing itself before it is sealed to the pumps. The mercury is rolled to the electrode farthest away from the tubulation. This practice can be followed only with special types of electrodes, and even then it is not an entirely satisfactory method of processing. During the high and intense heat treatment followed while bombarding, the mercury vapor will combine with the impurities released from the electrodes and the tubing and will cause darkening. It is not too satisfactory a process although it is recommended by some manufacturers. Such a method would be ideal if this blackening or darkening could be avoided, because the heat treatment during bombardment will produce a more uniform distribution of mercury than is possible with any other method of mercury insertion.

The two methods found most satisfactory for inserting mercury into a tube are the side-arm arrangement, or an injector method for which there are a number of devices on the market.

The Side-tube Method. In this method, a small side tube is sealed to the tubulation extending from the letter before the whole unit is attached to the manifold for pumping. This mercury trap, illustrated in Figure 10.4, contains a bulb for the mercury and a short length of tubulation glass which connects the bulb to the tube proper. The bulb is filled with the finest mercury available and sealed to the tube, bulb end down. It is important to keep the mercury out of the letter until the bombardment is over, so that no mercury is allowed to spill from the bulb into the tube. If it does so, and especially if the glass is not absolutely clean, the mercury will form little droplets on the side of the tube. Any mercury in the letter during bombardment may combine with impurities and form black spots which are difficult to remove or which may cause excessive darkening of fluorescent tubing. Also, if the mercury gets intô the tubulation, it may cut off the tube from the pumps. Therefore it is necessary to keep the mercury in the bulb during all sealing, bombarding, and pumping operations.

With the side tube in place, the letter is sealed to the manifold in the usual manner. The bombarding operation is completed, and the pumps are used to bring the pressure down to the best vacuum obtainable. The main pumping stopcock must be turned on slowly; otherwise a sudden intake may pull the mercury from the trap into the system. With the pumps still on, the mercury side tube may be heated gently with a soft flame to eliminate moisture. After the mercury trap has been heated, it is advisable to run the hand torch over the entire tubulation and over the glass leading to the mercury cup. This

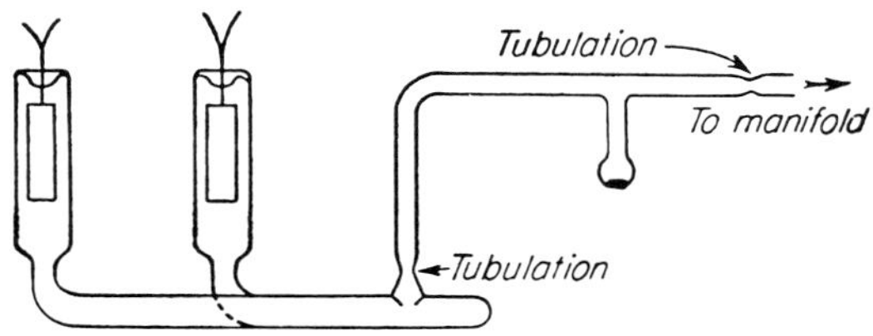

Figure 10.4. A mercury trap connected to a letter to be processed.

releases any moisture that may have condensed on the walls of the glass during the bombardment of the tube. When the proper vacuum has been obtained, and the tube is filled with rare gas to the proper pressure (see Table 6.1), it is then sealed off with the mercury arm attached. The whole letter is tipped in such manner so that the drop of mercury in the trap flows into the letter itself. The mercury trap is then sealed off from the letter. After the tubulation tip has cooled, the mercury can be distributed throughout the letter.

The Mercury-Injector Method. Although many injector methods are possible, the two described are most commonly used.

Figure 10.5 illustrates a manually operated mercury injector. It consists of two glass parts—a lead glass piece which seals to the manifold itself and a pyrex piece which fits into the lead glass by means of a ground-glass joint. The mercury is inserted into the ball on the end of the pyrex piece which is kept in a down position while the letter is being processed. When the mercury is ready to be inserted into the letter, the pyrex piece is rotated 180 degrees to permit the clean mercury to flow directly into the tubing.

The electric mercury injector is illustrated in Figure 10.6. This injector is also a fixed part of the manifold system. On the top is a reservoir of mercury which is kept under vacuum at the same time as the letter is being processed.

To insert the correct amount of mercury into the letter, a push button closes an electrical circuit through a magnet placed on top of the injector. The force of the magnet lifts an iron plunger with the correct amount of mercury which then drops into the tube when the magnetic circuit is opened.

The size of the slot in the plunger determines the amount of mercury to be inserted at one injection. Many drops of mercury, if necessary, can be inserted by opening and closing the magnetic circuit the required number of times.

Quantity of Mercury Necessary. The quantity of mercury to be inserted into any luminous tube depends upon a number of factors. However, it is to be noted that the minimum amount of mercury should be used.

Mercury-filled tubes can be divided into two classes: tubing for signs and cold cathode tubing.

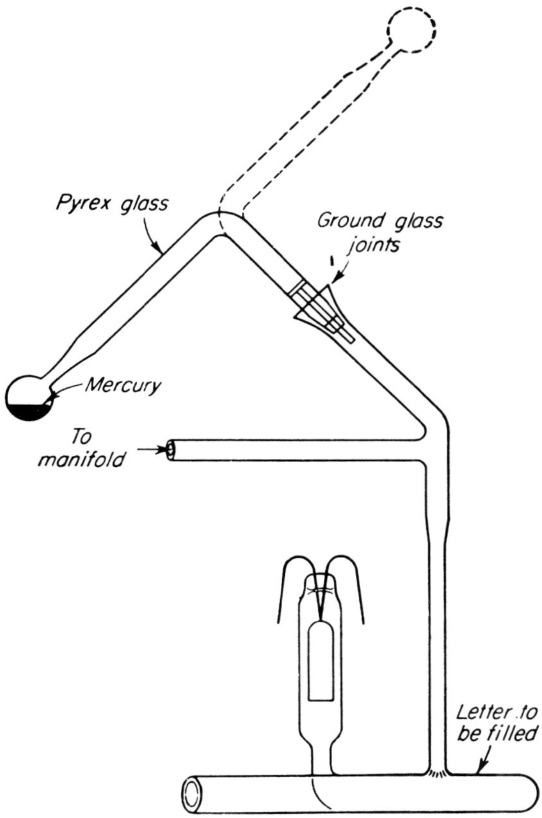

Figure 10.5. A manually operated mercury injector.

Tubing for signs can be subdivided as follows:

1. Long units—more than four feet long
2. Short units—less than four feet long
3. Tubing to be used in warm climates
4. Tubing to be used in cold climates

The amount of mercury to be used should be the minimum required to produce the necessary mercury vapor discharge. For long units and cold weather conditions a little more mercury is used. For short units or warm weather conditions less is used.

The following list should be used only as a guide for the quantity of mercury:

1. Long units—cold weather—600 milligrams

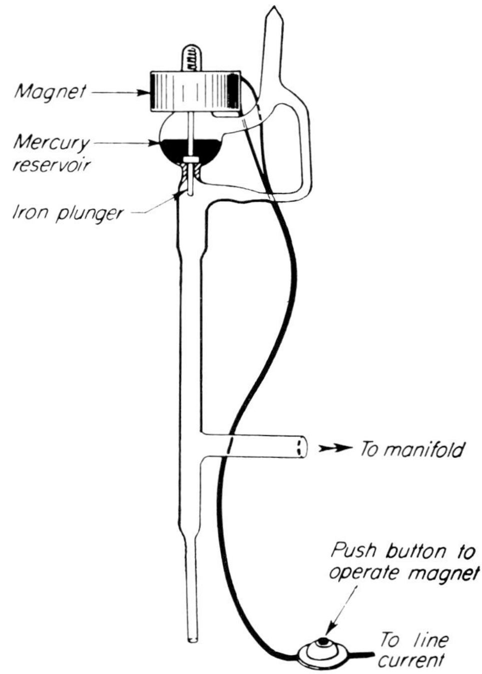

Figure 10.6. An electric mercury injector operated by a magnet lifting an iron plunger.

2. Long units—warm weather—400 milligrams
3. Short units—cold weather—400 milligrams
4. Short units—warm weather—300 milligrams

Cold cathode tubing, irrespective of diameter, should contain only a minute quantity of mercury. The following list can be used as a guide for the quantity of mercury for cold cathode tubing:

1. 15, 18 or 20-millimeter tubing used on 30- or 60-milliampere transformers—200 milligrams
2. 20, 22 or 25-millimeter tubing used on 100, 120 or 200-milliamperes—50 to 100 milligrams

If the manufacturer can obtain complete vaporization throughout the tubing with smaller quantities of mercury than indicated above, then the smaller amounts should be used. It must be remembered that the more mercury used in a tube, the greater the possibility of streaks, spots, blackening and dirty rings developing during tube operation. On the other hand, too little mercury may not give complete vaporization throughout the tubing, especially in long tubes used in signs in cold climates.

In cold climates where temperatures go below 32°F (0°C), a mercury-filled tube will lose brilliancy because the mercury will not vaporize. The quantity of mercury used is not a factor at these extremely low temperatures. The manufacturer should design the sign to use higher current transformers or higher resistance rare gas, or he may shield the tubing by using channel letters enclosed in front with plastic. In extreme cases he should eliminate mercury tubing entirely from the sign and use tubing that operates on rare gas only.

Aging Methods. A tube properly heated, with clean glass and electrodes and properly evacuated, should not require aging. But there are several reasons why aging is almost a universal process in the neon industry. One reason is that in tipping off the tubulation impurities are released from the glass. Another is that attempts to speed up production have sometimes shown that a shorter time on the pumps and a longer time on the aging table are the most economical way of doing things. In the small plant, where high speed operations are rarely possible or even desirable, this latter reason does not hold true. But for many reasons even carefully made tubing does not immediately come up to color, and it must therefore be aged.

Aging is carried out by operating the tube at rated current for a period of several minutes to two or three hours, although a well-constructed, properly bombarded tube should age properly in not more than five minutes. During the aging process, the impure gas in the tube cleans up by chemical action, while the inert rare gas, which will not combine chemically with the electrodes, remains free in the tube. Generally speaking, this clean-up process, being a chemical reaction, can be speeded up by running the tube hot during the aging period. For this reason, often a current somewhat higher than the operating current is used for aging the tube (generally a 60-milliampere transformer is used for sign tubing). This extra current increases the temperature of the tube and speeds the clean-up reactions. But too high a current has the undesired effect of freeing other impurities which would otherwise remain dormant in the electrodes and the glass of the tube. So a happy medium is required to speed up the aging process without overheating the tube to the point where new impurities are liberated. For very poor pumping systems, only rated current should be used for aging because, in this case, the tube is not required to stand more than the normal operating temperature, and the convenience of using a standard transformer more than makes up for the longer time required for aging.

The process of aging can be gauged roughly by watching the colors in the tube. On the first application of aging current to a neon-filled tube, the tube will probably be blue throughout its entirety or blue only near the electrodes if it is unusually well made. This blue color almost immediately disappears (unless the tube is a mercury tube) and is followed by a bluish red. Then the ends of the tube nearest the electrodes begin to come up to color, that is, to assume the characteristic brilliant red-orange glow of neon. The red glow surrounding the electrodes then spreads toward the center of the tube, rapidly at first and then more slowly.

If tubes fail to age up to color or run too hot there are several possible causes. The amount of impure gas may be so large that the electrodes cannot absorb the impurities rapidly enough, while these impure gases are being released from the glass or electrodes which are operating at a very hot temperature. If the vacuum system is known to be in good working order, a very slow leak at the electrodes or in the glass seals may be the cause. Or, the gas in the gas bottles may have become contaminated, either with air or with other forms of impure gas. This can usually be detected by applying the sparker to the rare gas containers. Poor electrodes may be the trouble, especially if they are not made of the very purest metal or have not been thoroughly processed, since they then will proceed to liberate impurities at a slow but sure rate which will continually interfere with the operation of the tube.

If a tube takes very long to age (more than two hours), it is usually wisest to give it up as a bad job and begin again. The tubing may be opened, cleaned, and used again provided that it contains no mercury and provided that the electrodes or electrode seals are not the cause of the trouble.

Figure 10.7 shows a cutout relay and wiring diagram. This device is very useful to prevent fires and shock to the operator. It is connected in the primary circuit when aging tubing and particularly when kept burning overnight. The relay is designed for a definite narrow range of current. If the current falls in the primary circuit owing to a defective tube in the secondary circuit, then the magnet bar carrying the contacts is released and the primary circuit is opened.

The Completed Tube. When the tube has been properly aged, it is ready for installation in the sign itself. The methods used for mounting the tube in the sign depend largely on the type of sign mounting, that is, whether the sign is a box sign, a skeleton sign, a roof structure, a border sign, a window display, or a window border. Each of these types of signs (see Chapter 5) requires a different type of mounting and a different method for handling the completed tube.

Painting Out Crossovers. Before the tubing section can be said to be complete, the crossover connections between letters must be blocked out so that they do not detract from the clear-cut appearance of the sign. This blocking can be accomplished satisfactorily in a variety of ways. Painting is the most satisfactory method, but often a winding of tape is used to cover the crossovers.

Whatever method is used, the primary requirement is permanence. Nothing destroys the appearance of a sign more quickly than half-covered crossovers; and if a customer complains of this fault before the guarantee is up, he has every right to service. So only the best materials should be used for this purpose.

There is one rule to observe if paint is used: Never use paint with a metallic base (lead paint, a copper base paint, etc.). Metallic paints will conduct electricity. If they are used, particularly if they are used near the electrodes or near a sharp bend on the tube, they are bound to cause trouble. If a voltage is built up between the paint and the nearby box housing, as

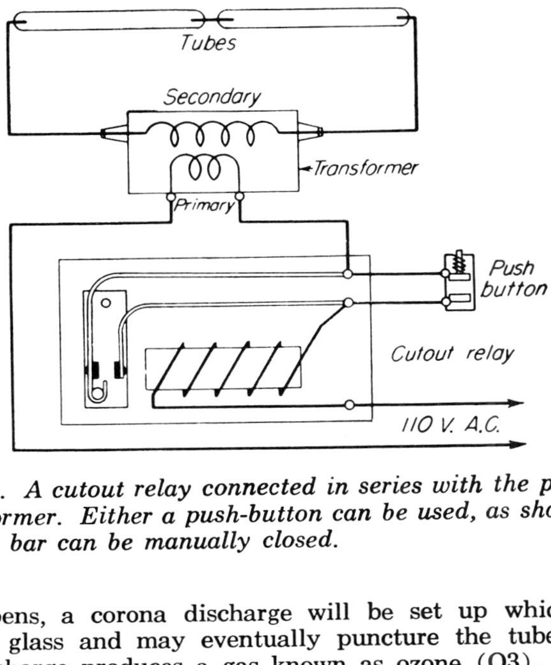

Figure 10.7. A cutout relay connected in series with the primary of a transformer. Either a push-button can be used, as shown, or the magnet bar can be manually closed.

often happens, a corona discharge will be set up which will attack the glass and may eventually puncture the tube. This corona discharge produces a gas known as ozone (O3), a form of oxygen molecule which has a very destructive effect on paints. Block-out paints should have a resistance to this type of attach. Metallic paint may also cause considerable radio interference when the voltage on it discharges.

Non-metallic paints for tube painting have been specially developed and can be obtained either from a sign supply house or from a reputable paint manufacturer. These paints should be thoroughly opaque when applied in a single coat. Sometimes two coats are given to ensure long wear. The paints should also have a high degree of adherence to a glass surface, a quality which all non-metallic paints do not possess.

The paint generally accepted for this use is an enamel, especially made to have high insulating properties as well as to resist the weather, heat, light, high voltage, and corona discharge. Any large paint manufacturer, especially one catering to the sign trade, will be able to supply such a paint at a reasonable price.

Block-Out Tubing. Another method used to block out the glass between letters is by the use of a tubing called Blok-Out tubing. Blok-Out tubing is made of regular clear glass sign tubing which is specially treated by the manufacturer to produce a black deposit that is near the surface but actually within the glass. Blok-Out tubing is black in color and, therefore, opaque, preventing any light from coming through.

The advantage of Blok-Out tubing is its permanent black-out quality. It is not affected by weather conditions. On the other hand, the cost of Blok-Out tubing is relatively high and also the cost of manufacturing the tubing is higher because of the labor involved in inserting pieces of Blok-Out glass. This method is ideal where cost is not a factor and where a high quality permanent installation is essential.

CHAPTER XI

The Complete Luminous-Tube Sign

Bending the glass tubing, bombarding, pumping, and filling the tube is only part of the process in the manufacture of luminous tube signs. The completed glass tubing must then be properly mounted and connected electrically to the various components.

Many methods are commonly used throughout the world in completing the finished luminous tube installation. The two principal methods can be defined as follows:

1. Sign completely assembled in the shop.
2. Sign assembled on the job.

In the first category the sign manufacturer makes the complete box, properly painted, mounts the tubing on the face or faces of the sign with elevation posts, insulates the electrodes with electrode receptacles, places the transformers inside the box, and makes all the connections for low voltage and high voltage. The sign is then ready to be transported to its destination and hung. The only work required besides hanging the sign is to make the necessary low voltage connections from the primary switch in the building to the low voltage wires of the sign. In most cases the sign is transported with all its tubing attached, and very little breakage is encountered by experienced installers. Where the sign is large and the tubing complicated, the sign can be installed without the tubing, which is then put in place after the sign is installed.

Also in the category of signs assembled in the shop can be included many types of skeleton signs which are installed as one unit, that is, with the transformer and the tubing mounted and in place ready for installation. The skeleton tubing may be tied together with frames of glass, or special aluminum frames that can be purchased especially for quantity production jobs.

The second category, signs assembled at the job, includes both outdoor signs and skeleton or border signs hung indoors. Most of the work of installing and fitting is done on the job itself rather than in the sign plant. In many countries, the only completed items taken to the job may be the tubing and cut-out metal letters. All the other necessary work, such as the metalwork, soldering the metal letters in place, cutting up angle iron, making the necessary connections, and installing the transformer itself, is done by the workmen at the job site. This method involves working under poor conditions, with the result that labor expense is much higher than if a sign were completely finished in the plant.

In the case of border outline tubing, either in windows or on outdoor installations, completion of the work at the job cannot be avoided.

Shop-assembled Sign. The fundamental mounting part of the glass tubing is the elevation post or tube support. The elevation post is correctly located and screwed down in place on the face of the sign cabinet. This tube support usually comes in two parts, the metal to be fastened to the sign and the glass part fastened to the tubing. The tube support is adjustable both in height and in location. When the glass is fastened to the tubing by means of fine tie wire, the support holds the tube in place, provides the pressure with which the electrode lead wire presses against the spring in the electrode housing, and in general maintains the tubing in its correct position. It is important to avoid straining the sign glass when mounting the tubing on elevation posts. A strain may cause the tubing to crack while it is being installed or afterward when the tubing is in position and operating.

Many different types of tube supports are available, varying in height from one to six inches.

A box sign may use either complete electrode receptacles, such as housings, or ring bushings. An electrode housing is usually constructed with a compression spring which makes contact with the electrode terminals when the tubing is installed on the face of the sign. The electrode housing may use a pigtail wire instead of a spring. The pigtail lead is connected to the wires of the electrode before the tubing is put in position. The spring arrangement is more popular because it permits the installation of tubing without extra work in making connections.

Ring bushings can also be used instead of housings. The ring bushing merely insulates the glass electrode from the metal box. Necessary connections have to be made on the inside of the box from the electrode wires to the high voltage cable.

A box sign requires the use of cable supports placed on the inside of the sign in such locations that they will firmly hold the high voltage cable. With electrode housings, these cable supports can be located farther away from these receptacles because the housing itself acts as some support for the cable. With a ring bushing, the cable support must be close to the bushing in order to support the cable and prevent the electrode connection to the electrode from touching the metal. Cable supports must be correctly spaced to give maximum support and insulation to the high voltage cable that is connected to the transformer.

Installation of Transformers. Many transformer failures may be caused by water getting into the transformer casing and coming in contact with the windings. For this reason, the sign box should be constructed carefully, with tight seams and with as close a fit as possible around the electrode housings. Drain holes should be drilled in the base of the metal sign box so that any water which collects may drain off. If the transformer has to be mounted on the bottom of the box, it should rest on channel iron runners so that drainage water will run under it. The transformer should also be mounted in that location which allows for short high-voltage leads from the tubing.

Skeleton-type Signs and Mountings. The skeleton type of sign is usually self-supporting; that is, it is not solidly fastened by means of elevation posts. A small sign of this type may hang directly from its electrodes without further support, although it is always advisable to have extra wire supports from the ceiling to the glasswork itself. Larger hanging skeletons may be supported on several wires, as shown in Figure 11.1, with the transformer mounted above, below, or to one side of the windows. In large skeleton signs, the tube is made up of several sections which must be connected together. The break in the sign between two tube sections should come between letters, of course. If it is a script letter sign, no break at all should be visible. In this case, specially designed connections, shown in Figure 11.2(*c*), are used to hide the break.

The separate sections of a skeleton sign must be strongly connected, both electrically and mechanically. The usual practice is to slip a piece of glass tubing over the first electrode, then connect the electrode wires together, and then slip the outer tube back so that it covers both electrodes. This style of connection is neat, electrically safe, and it harmonizes with the remainder of the sign tubing. With hanging transformers, a long glass tube is used to insulate both the cable and the electrode connection. Approved metal connectors bind the cable to the electrode wires.

Types of Transformer Mountings for Skeleton Signs. There are three ways of mounting a transformer used with a skeleton sign in an inside location:

1. A transformer can be of the hanging transformer type which is already provided with high-voltage leads, with hanging brackets, and also, in some cases, with a pull-chain switch.

2. In some cases, a hanging transformer is not desirable. A standard transformer can be used, mounted in a metal box and placed at the most convenient point either above the skeleton sign itself or to one side. In all cases the high voltage leads must be protected.

3. Some transformers for skeleton signs come equipped with porcelain housings. A transformer of this sort can be either placed on a ceiling with the electrodes from the skeleton sign going into the transformer or placed on the window sill with the skeleton sign coming up from it.

Skeleton signs can be classified as follows:

1. A simple skeleton sign of one color.
2. A skeleton sign which has two or more colors and possibly more than one line of reading matter.
3. A skeleton sign which has one or two colors and a border.

If a skeleton sign contains more than one line and more than one color, it becomes quite complicated, and it is very necessary that it be thoroughly laid out for the glass blower so that the

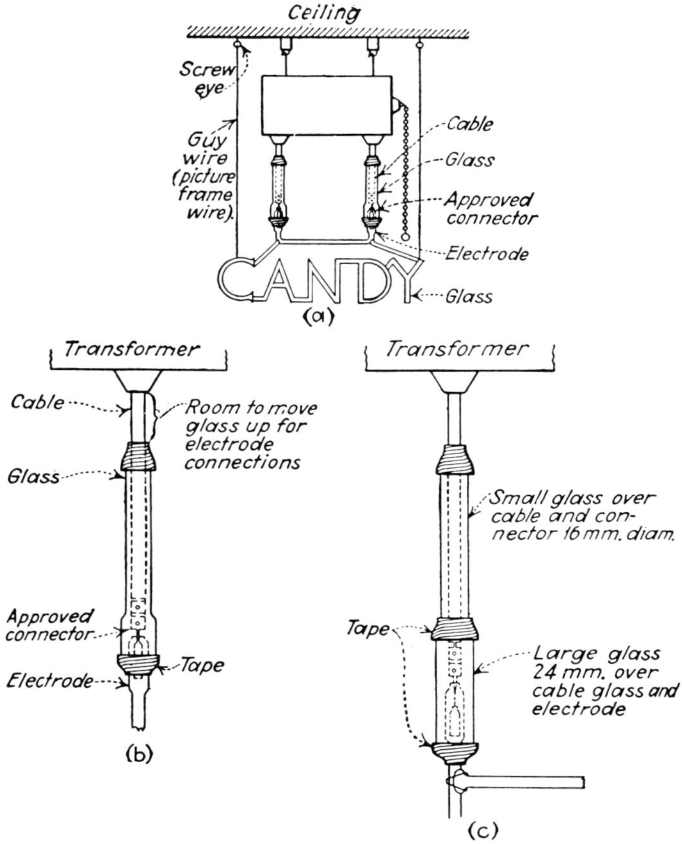

Figure 11.1. (a) Hanging transformer and skeleton sign. The guy wires should be completely insulated when attached to the neon tubing. Note the method of protecting electrodes and cable with glass tubing. (b) Detail of glass sleeve for protecting electrode and cable. (c) Two pieces of tubing used in place of the one shown in (b).

points of connection between the two sections are as short as possible and not visible to spoil the appearance of the whole sign.

Border Mountings. Tubes used for building outlines or for window borders are usually supported on elevation posts [Figure 11.2(*a*)] since this is the most permanent and foolproof mounting. But occasionally window borders are hung from wire supports. If elevation posts are used, they are fastened either to the window trim or to a built-up metal or wood framework. Elevation posts for window borders may vary in height from two to six inches. Special extensions may be used in conjunction with the standard post to give the necessary height. Hollow flares can be made from tubing in any height desired [Figure 11.2(b)]. The tube is then bound with wire to the posts, in the same manner as in the box-type sign. The separate sections of border-type tubes must be fastened together in the same fashion as the skeleton signs, but in this case it is doubly important not to allow a break in the sign to interfere with the continuous line of the sign. The bending back of the tubing so that the electrodes are hidden behind the sign, especially if the break is offset, will prevent any break in the sign from being visible to the public.

The glass sleeve which insulates and protects the connections between sign sections should be used for border signs as well as skeleton installations.

High Voltage Wiring. General Practice. Regardless of whether the sign is a box type, skeleton, or border, there are a few general rules which should be followed in all high voltage wiring. High voltage wiring should be kept as straight and as short as possible. Sharp corners produce a mechanical strain that may affect the

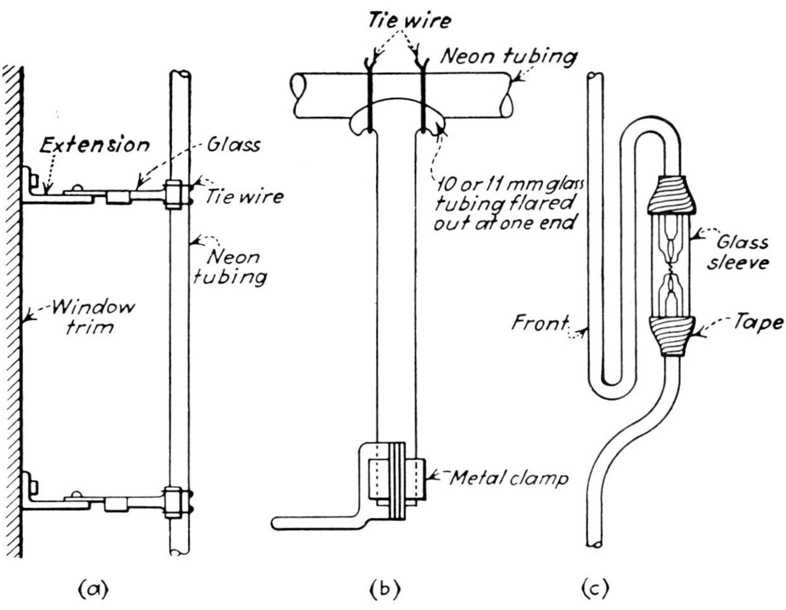

Figure 11.2. (a). Extension posts for window border signs. (b) Extension post made from glass tubing and mounted with metal clamp. (c) Method of connecting electrode of two tubing units in a window border. No break is visible to the public.

insulation. Furthermore, the high voltage wiring should be kept at least $2\frac{1}{2}$ inches away from the nearby metal box of the sign or any other metal parts.

Connections to the transformer should always be made with high voltage cable especially designed for neon signs. Cable should never be used unless it is perfect mechanically and electrically, and it should carry the approval of Underwriters' Laboratories or similar organization for the voltage used. When one cable crosses another, there should be ample spacing between the two or glass tubing should be used over each cable to insulate one from the other.

In summary, the requirements for the high voltage circuit are (1) a straight, continuous series circuit from one terminal to the other; (2) good metallic contact throughout the circuit; (3) well-insulated cable in perfect condition; (4) strong electrode connections, preferably with approved electrode housings; (5) complete protection from the weather. In wiring the signs, all wires should be kept as short and as straight as possible, and all metallic parts of the circuit should be protected from the metal parts of the sign cabinet and its fittings. Never use any other material but brass or copper for high voltage electrical connections.

The Low Voltage Circuit. The power available in the United States is usually 110 volts or 220 volts, 60 cycles. In some locations it may be either 25 or 50 cycles. In many other countries both the voltage and frequency may vary from standard practice in this country. It is therefore very important to determine both the voltage and frequency available at the place where the sign is to be installed. The transformer then can be ordered with the primary winding in accordance with the voltage and frequency required.

In a great many localities, the voltage at the sign may be low. Instead of 110 to 115, it may be as low as 100 volts. Under these conditions, if the transformers have been loaded with the maximum amount of tubing, the sign may flicker. For this reason, it is always advisable to test the voltage before designing the sign, using a standard alternating current voltmeter, of 0 to 250 volts range. If the voltage is below standard, allowance must be made in the layout of the tubing. In addition, the leads from the primary outlet to the transformer should be kept short and should be made of as large wire as possible, so that the voltage will not drop to any great extent between power source and sign.

Transformers of 7,500 volts and over usually have the mid-point of the secondary winding grounded to the metal case or a separate lead brought out which can be connected to a ground. If a transformer is not installed in the metal box of the sign itself, but at some other point which insulates the case from the ground, then a wire should be connected from the metal case directly to the ground. Otherwise an electrical strain may be produced between the tubing and the metal sign, which may cause punctures, buzzing, or interference in radio sets nearby. It is always important to make sure that a good ground connection to the transformer is made.

Whenever possible, the sign should be connected to the electric mains through a separate circuit, not connected to any other appliance or electrical device. If other loads are taken from the sign supply circuit, the voltage may drop to the flicker point when the other loads are switched on.

When a metal box-type sign is installed outdoors, usually a special metal outlet box on the outside of the building is placed at a convenient point near the sign. Approved BX cable connects this outlet box to a similar outlet box on the sign itself.

All low voltage wiring should be made in accordance with the regulations existing in the locality. The type of wire, connections, and outlet boxes should conform to both Underwriters' Laboratories specifications and any other requirements which the local authorities specify.

Multitransformer Installation. The wiring diagram of a three-transformer sign is given in Figure 11.3. It will be seen that the high voltage circuits are kept isolated from one another, electrically at least, each one being connected to its own load of tubing. The primary circuits and switching devices, if there are any, are connected together at the power or low voltage side. For this reason, a single master switch, which connects all the primaries to the power line, is used. If there are many transformers in the circuit, the power consumed may be quite high. It is important to determine whether the low voltage lines available will be able to carry the necessary power for a large installation. In a large installation it may be better to use a three-wire power line to the sign with the distribution equally divided between the two circuits.

Correct Loading of Transformers. When the transformer for a sign is selected, it is assumed that the sign will be correctly loaded. The transformer is chosen by reference to the data given in Table 6.2, according to the number of feet of tubing, the diameter of the tubing, and the gas to be used. When the sign is actually built, it is assumed that the transformer will operate as it should without further attention. Fortunately, in the great majority of cases, this procedure is a safe one to follow. But in some cases, the transformer may be badly unbalanced. If it is underloaded, the transformer is not being used efficiently. If, on the other hand, it is being overloaded, two difficulties may arise. The transformer will run hot; this may eventually cause trouble. The most likely trouble is flickering. When a transformer is used to light a tube longer than it can safely handle, a radio frequency voltage is set up in its windings. This voltage may help to cause a transformer burnout. It may cause radio interference. It will cause the sign to flicker, especially in wet weather. All these results are undesirable. To avoid them it is necessary to load the transformer correctly.

When flashing signs of the high voltage type are used, the lengths of tubing connected to the transformer are not always equal; usually the length of tubing actually connected is increased

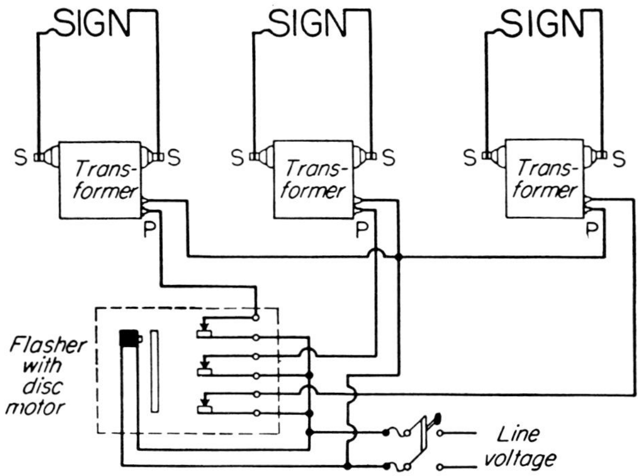

Figure 11.3. Multitransformer connection diagram, showing three transformers with a low-voltage flasher in the primary circuit.

as the spelling actions proceed. There is no sure way to avoid difficulties in this case, except to make sure that the transformer will safely carry the maximum amount of tubing to be connected at any time. The voltage drop across the flasher must also be considered.

All the possibilities above may be taken as sufficient reason for testing the loading of the transformer when it is actually in the sign. Although the transformer-loading test is not carried out by the majority of neon manufacturers, it is well worthwhile. There are a number of methods for testing transformer-loading conditions, such as, the use of a milliammeter in series with the secondary, the use of an ammeter in the primary circuit, the use of a wattmeter indicating wattage which is plotted against tube loading, and the flicker voltage test. In May, 1959, the transformer manufacturers changed the design of the luminous tube transformers so that they would now operate on an input voltage of 120 volts instead of 115 volts. Because of the fact that standard complete units with variable voltage auto-transformer types are available (Figure 11.4), the flicker-voltage test has been adopted by practically all neon sign shops for testing loading conditions. Transformers with a 240-volt primary should use figures double the left-hand column table below. These will result for that type of transformer with the correct results shown in the right-hand column.

By following the method to be described, the test is carried out by reducing the voltage applied to the primary of the transformer in gradual steps until the neon tubing begins to flicker. At this stage the voltage is noted. If the transformer is for 110-volt operation, it should be possible to reduce the primary voltage to 85 or 90 volts before flickering occurs. If this can be done, the transformer is being operated with a sufficient margin of safety.

These rules for testing the loading of transformers are given not only to check the size of the transformer as chosen by the use of Table 6.1, page 89, but also to allow the proper size of transformer to be chosen when the conditions given in the table are not actually met. Differences in gas pressures used and in the diameters of tubing will produce tubing which requires transformers of larger or smaller size than that indicated by the table. In this case, the only safe way to proceed is to measure several different types of transformers on the actual completed sign and to choose for permanent installation the one which suits the job best. In the great majority of cases, the transformer can be chosen by reference to the table, but any special case requires that the above tests be made.

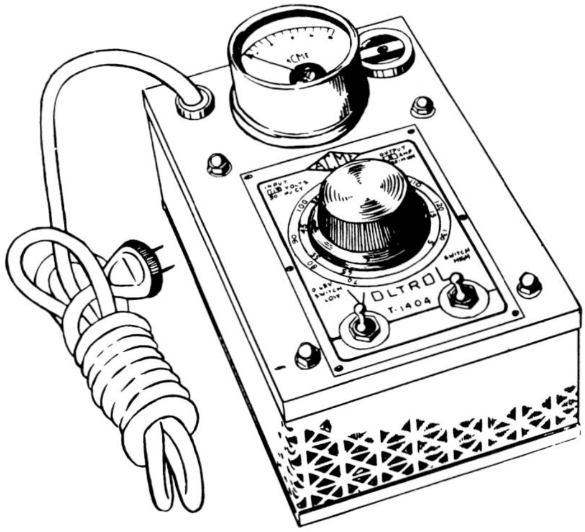

Figure 11.4. Tester for obtaining correct tubing load on a neon transformer. This device is inserted between the primary circuit and the transformer.

In many locations where transformers may be placed at a distance from the tubing, long cable leads may be required for connecting the end electrodes to the transformer. In some localities exposed high voltage cable is not permitted. In this case the cable must be enclosed in grounded metal conduit. This may give rise to corona difficulties, especially when transformers of 7,500 volts or more are used. Corona has the effect of causing a chemical decomposition of the insulation of the cable, with the result that in time the cable conductor may cause a short circuit to the conduit. Long lengths of high voltage cable in conduit should be avoided.

Besides cable damage, burnouts of the transformer windings may result. Cable in a metal casing builds up an electrostatic capacity to ground, which partly neutralizes the high flux leakage of the transformer. As a result, the terminal voltage of the transformer, even under load, rises to values much higher than normal. High voltage of abnormal values may puncture the insulation between the secondary windings. Although metallic sheathing may be required by local authorities and is satisfactory from a mechanical point of view, it should be employed only with short lengths. If possible, it should be entirely avoided. High voltage cable using polyethelene insulation shows great promise in overcoming breakdown in conduit.

Installing Rotary Converters. When a sign is to be installed on a direct current power system, a rotary converter is used (as explained in Chapter 4). The installation of these machines is fairly simple. The machine should be mounted preferably in the basement or in some other out-of-the-way location. This will deaden the slight noise which these machines produce when operating. The machine should be solidly mounted, but a layer of rubber under the wooden or metal base may be used to lessen the noise from vibration.

Two switches are usually employed, one for connecting the direct current to the converter, which is always turned on first, and one for connecting the transformer and sign to the output of the converter after it has had a chance to get up to speed. If the switch to the sign is closed first, the machine will start slowly, since it is then under load, and it may not come up to speed for a long time. During this time, the sign will flicker badly, and the transformer or rotary converter or both may be injured. It is possible to combine both switches into a single unit, so that the alternating current switch cannot be closed until after the direct current switch is on. Full operating instructions and directions for oiling the converter should be left with the owner of the sign.

Before installing, the direct current supply voltage should be tested with a test lamp or voltmeter to make sure that it is the proper value for the converter. A test of the alternating current voltage on the output side of the converter may also be made, but usually, if the direct current voltage is correct, the speed of the converter will be correct, and this will ensure the proper value of frequency and alternating current voltage on the output side. Both the correct alternating current voltage and correct frequency are of great importance. If the converter runs slow, even though the direct current voltage is correct, the frequency will be below the standard 60-cycle value, and the transformer may be burned out as a result.

Rotary converters built for neon signs are rated in volt-amperes. It is very important to match the volt-ampere reading of the converter with the transformer or transformers used. Converters rated in watts will arc at the brushes and eventually burn out if used with a transformer which, although rated the same as or lower than the converter in watts, has a higher volt-ampere rating.

To get the maximum economy from converters, it is advisable to use transformers of the high-power-factor type, that is, of 90 per cent power factor. In this case, a 15,000-volt, 30-milliampere transformer which has a 450-volt-ampere rating at 50 per cent power factor will have only a 250-volt-ampere rating at 90 per cent power factor. With a standard transformer, a 500-volt-ampere converter must be used; with a high-power-factor transformer, a 250-volt-ampere converter can be used.

Packing of Luminous Tubing. With the widespread use of luminous tubing for advertising and interior lighting, the manufacturers have been faced with serious problems pertaining to efficient shipment of the tubing.

The manufacturer must produce an article which can be shipped economically and safely. Therefore, in designing the sign, reasonable care must be taken to see that the sign may be safely shipped with a reasonable amount of packing. The layout man should consult with the shipping department as to the practicability of shipping the sign which he is designing. In practically all cases, the slight design changes necessary to ensure safe and economical packing of the sign will in no way detract from the beauty or effectiveness of the finished article.

Reinforcement of Tubing. The Freight Container Bureau of the Association of American Railroads recommends that all skeleton neon signs should be reinforced by glass tubing, wired and varnished to the sign itself, particularly at joints or where there are long, unsupported spans of lettering.

On large outdoor signs consisting of tubing mounted upon a rigid steel framework, it is recommended that the tubing be shipped separate from the steel background to lessen the hazard of breakage due to sudden jars or shocks.

When separate shipment is impractical, the tubing must be securely wired to the background. Where elevation posts are used, these must be secured to the background by steel clips. The posts must be positioned exactly so that no strains are set up in the glass, even when the tubing is tightly wired to the background.

It is likewise important that the terminal hole be of sufficient diameter so that the tubing will not come in contact with the sides of the hole. A live rubber bushing should be used between glass and either metal or porcelain surfaces to retain the positioning and absorb vibrations.

Tubing coated on the inside with fluorescent powder must be joined and shaped with great care by the glass blower. During shipment of a sign, the powder may break away from the glass of a carelessly processed tube or the glass may crack at a joint or at a weak bend.

Crating the Neon Sign. The glass tubing comprising the letters and border of the sign must be rigidly fastened to either the background of the sign itself or to a rigid wooden frame; then either the frame or the whole sign must be cushioned in the crate against vertical and horizontal shocks.

The use of sponge rubber for cushioning neon signs is economical and effective and may be used with practically any shipping container. Cushioning pads should be of ½-inch thick sponge rubber placed between the tubing and the wooden frame members. Soft copper wires should be used to tie the tubing in place, and the spacing of the pads should be such that there are no unsupported spans greater than 12 inches in length. A pad should be placed between the glass and frame at every place where the tubing crosses a wooden member.

The rigid suspension frames carrying the glass tubing must likewise be cushioned in the shipping container. If sponge rubber pads are used for this purpose, they must be at least one inch thick when placed at the bottom of the container and at least ½ inch thick when used to cushion the top or ends of the frame.

Commercially manufactured rubber bands may be used for this purpose, but rubber bands cut from automobile inner tubes should never be considered. The bands must be fresh and made from live rubber. The bands must be kept in place under slight tension, and at least three inches clearance must be allowed on all sides between frame and container. This method of cushioning the frame is particularly well suited for use by the small manufacturer.

Large outdoor signs must be cushioned by the use of steel compression springs. The use of steel springs requires a fairly good matching of spring tensions and the use of well made springs. As in the case of rubber band suspension, the steel springs should be maintained under slight tension and a three-inch clearance allowed.

Types of Shipping Containers. In the design of nailed wooden crates, the three-way corner type is recommended. At least one face of the crate should be secured with screws or double-headed nails to permit the customer to open the crate without damage to the sign. If double-headed nails are used, provision must be made to prevent damage from the protruding heads. Reuseable crates are often used to ship signs between factories and distribution points and for shipments in which the sign is to be repacked for storage or shipment. It is essential here that opening and closing be done without damage to either the container or its interior blocking.

Wire-wound and other prefabricated crates make satisfactory shipping containers for neon signs. When adequate cushioning is provided to keep the sign in position, the strength and light weight of these crates are exceptionally advantageous.

Corrugated fiberboard boxes have been used successfully to ship neon signs of the small indoor variety. The corrugated boxes offer some cushioning in themselves, but this is not adequate, and interior cushioning is necessary as in the case of the wooden box. In this connection, excelsior and corrugated fiberboard pads should never be considered adequate cushioning.

One outstanding disadvantage in the use of the corrugated

fiberboard box is the fact that the freight handlers are unable to see the contents of the box and consequently will be unaware of its fragile nature. For this reason, a standard shipping label has been adopted throughout the luminous tube industry.

Large outdoor signs may be packed and shipped using the methods described above. In the case of these large displays, open wooden crates are universally used as shipping containers.

CHAPTER XII

Flashers and Time Switches

It is an axiom in the sign business that a good sign can be made better by the addition of animation, that is, by flashing sections of the sign in some attention-compelling sequence. Even the simplest unit can be made more effective if it is flashed on and off. More elaborate signs gain if the sign is "spelled out" or if units in the design are lighted in sequence. For this reason the subject of sign flashing is a most important one, one which the sign manufacturer should understand thoroughly if he is to offer his customers a well-rounded service.

Primary vs. Secondary Flashing. There are two general types of flashing applicable to luminous tube lighting. *Low voltage flashing* is accomplished in the primary circuit of the high voltage transformer. This system has the advantage of operating at low voltage; so the switching mechanism can be rugged and of simple construction. Moreover, the flasher may be located remotely from the sign, where it is easily accessible and readily maintained and serviced. This is the type of flasher universally employed for simple "on and off" flashing of a sign as a whole, or for flashing sections of a large sign. The disadvantage is that a separate transformer is required for each flashing section of the sign. The *secondary* flasher, on the other hand, is used to switch a single transformer to two or more sections of a sign. Therefore, one transformer can be used to animate many sections of tubing. The disadvantages of the secondary type of switching are the necessity of running long high voltage cables from the flasher to each section of tubing and the difficulty of maintaining a high voltage flasher in good repair because of the high voltage arcing at the contacts. Several satisfactory types of high voltage secondary switches have been developed.

The type of flasher to be used depends on the size of the sign, the number of sections to be flashed, and the overall cost of the installation. Each installation must be analyzed individually before designing the switching circuits.

Thermostatic Primary Switches. The simplest type of primary switch is the thermostatic blinker used for simple on and off switching of a complete sign. The connections of this type of switch are shown in Figure 12.1. In series with the transformer primary is connected a coil of resistance wire, wound around a bimetallic strip. As the current passes through the heater coil, the coil heats the bimetallic strip. This causes the metal to bend and snap, opening a pair of contacts in series with the heater coil. This breaks the circuit, turns off the sign, and allows the bimetallic strip to cool. When the strip has cooled sufficiently, the metal snaps back to its original form and the contacts are again closed, the sign lights, and the heating begins anew. Thereafter the cycle of operations repeats itself, and the sign flashes on and off at a more or less regular rate.

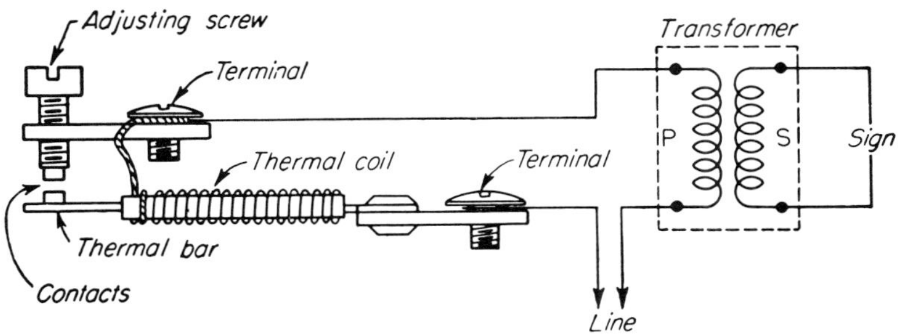

Figure 12.1. A thermostatic type of flasher.

The thermal type of flasher is simple and inexpensive, but it has several distinct limitations. The contacts are not large enough to carry heavy loads. Thus the transformer load connected to a typical thermostatic flasher must not exceed 100 volt-amperes. The thermal type of switch is made in a variety of forms, from a small unit which may be fitted directly into a lamp socket, to a larger structure resembling a relay. The latter type of unit may be fitted with more than one contact and may throw in two directions.

The thermal type of flasher has not been adopted by the sign industry. It was used for small incandescent lamp display signs put into windows. The timing of the flasher is apt to be very irregular and varies widely with the temperature as well as with the size of load connected in series with its heating coil.

Motor Driven Primary Flashers. The type of primary flasher used whenever precise timing is required is the motor driven type. In one form, much used for incandescent signs, the motor is geared or belted to a rotating drum. Brushes make contact with metal segments on the drum. A more modern form consists of a motor driven camshaft which actuates pairs of contacts.

The drum and brush flasher is capable of handling loads up to 1000 watts per brush. Highly elaborate units may be built up from combinations of drums, with as many as 50 brushes. The drum is made up of metallic segments, so arranged that they travel under the brushes, making contact during a portion of each revolution. When special animation effects are desired, it is customary to have the drum manufactured for the particular job. At one end of the drum is a continuous segment, called the collector ring, which makes continuous contact with its associated brush. This establishes the connection to the other segments.

The camshaft type of rotary flasher is illustrated in Figure 12.2. Riding on each cam is a set of silver contacts which are opened and closed by the pressure of the cam as it rotates. Loads up to 3000 volt-amperes per contact may be handled with this type of flasher, and for this reason it is used for practically all neon sign applications.

For standard flashers carrying a small number of circuits, a shaded-pole induction motor is used. This type of motor operates

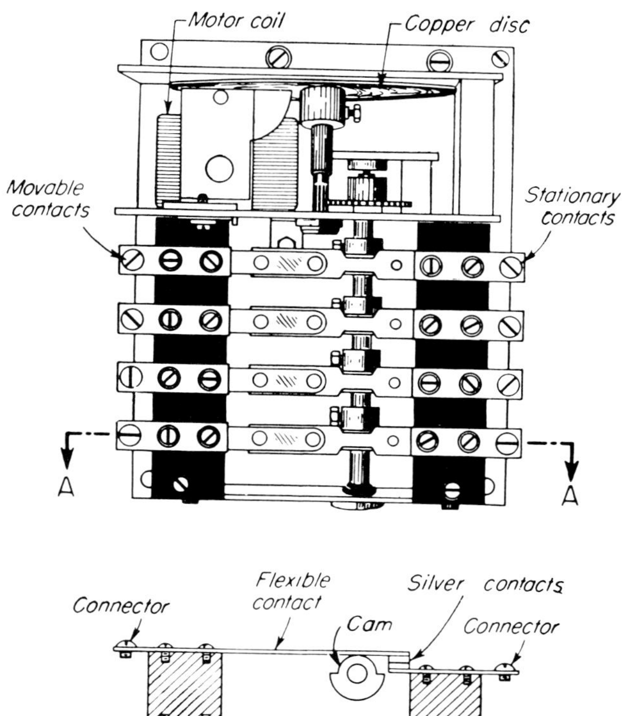

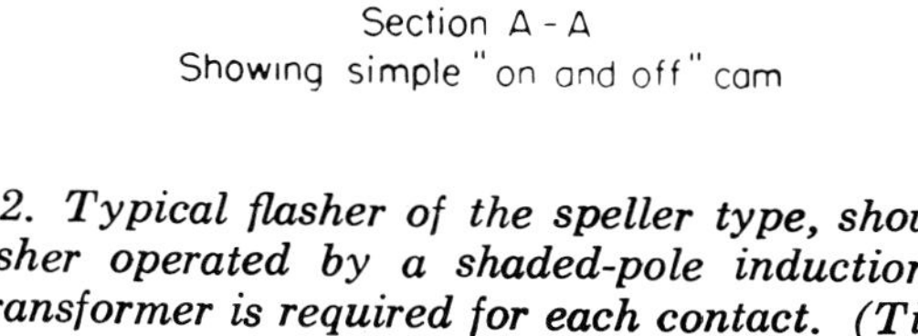

Section A - A
Showing simple "on and off" cam

Figure 12.2. Typical flasher of the speller type, showing a four-circuit flasher operated by a shaded-pole induction motor. A separate transformer is required for each contact. (Time-O-Matic Company.)

normally at a low speed, so that a simple gearing-down mechanism suffices. For a large number of circuits, a wound-rotor type of induction motor may be used. This type of motor has the necessary power to handle a large number of cams.

The cams in a flasher may be adjusted to give a wide variety of timing cycles. As a rule, the type of flasher required comes provided with the desired sequence of flashing. The cams are already cut at the factory to give the effect. The installation man may elect to make an adjustment himself by cutting the cam for his required timing cycle.

The standard types of flashers that are produced for low voltage operation are known as (1) on and off, or alternate; (2) speller type; (3) high speed border chaser. Any of these types can be combined with each other to give the desired flashing effect. The on and off, or alternate, type can be provided with either single or multiple contacts to give simple on and off flashing effects or an alternate flashing effect. The alternating flashing effect is obtained by two circuits alternating with each other. The speller type flasher produces a spelling action as one circuit after another is lit in sequence. The high speed border chaser flasher (Figure 12.3) is usually used with incandescent lamps and can be made to produce either a running border effect or a twinkling effect, depending upon the wiring of the circuits. The high speed border chaser type can also be used for neon tubing.

Either the on and off, or alternate, flasher, or the speller type flasher can be obtained in a variety of flashing speeds. The simple on and off flasher usually operates at a speed of 15 to 30 flashes per minute. By changing the gearing arrangement at the factory, or by cutting the cams differently, the speed can be varied from the standard. Standard speller flashers can be obtained in a variety of flashing speeds, from a fast speed of 2 to 10 seconds per cycle, to slower speeds of 4 to 20 seconds, or even 8 to 40 seconds per cycle. The two numbers such as 4 to 20 mean that the signman can adjust the speed of the flasher so that it takes anywhere from 4 to 20 seconds to complete a cycle. The high-speed border chaser can be adjusted to give 150 to 430 flashes per minute.

When applying a flasher to a particular job, it is necessary to operate the contacts so that they carry no more than the rated load in watts or volt-amperes, as specified by the manufacturer. Failure to follow the ratings may lead to short contact life and consequently increased maintenance costs. It also may result in faulty flashing as evidenced by the irregular operation of the flasher and may be accompanied by heavy arcing and radio interference.

Secondary Flashers. The Rotating-Gap Type. The primary flashers just described are not applicable to the vast majority of signs in which the cost of a separate transformer for each section cannot be justified. The need for a reliable secondary circuit flasher is thus apparent. One type of secondary flasher which has

Figure 12.3. Four-circuit high-speed border-chaser flasher. Cam revolving in center opens one circuit at a time. This flasher is used to obtain running border or scintillating effects.

proved satisfactory for long trouble free service is the rotating-gap type. One form of this type is shown in Figure 12.4. The separate contacts are arranged in a circle at the top of the device, usually four to eight in number. The common contact rotates directly underneath, driven by a shaded-pole induction motor and a simple gear-down drive. The rotating element does not make direct metal-to-metal contact with the fixed elements, but rather passes beneath them, leaving a small air gap between. The high voltage from the transformer secondary is sufficient to break this gap, and a small spark is formed which carries the transformer secondary current to each of the terminals in succession. The lack of direct metallic contact between the operating surfaces increases the life of the device, without introducing any serious loss in the secondary circuit.

The speed of rotation of the motor may be controlled over wide limits, five to 35 revolutions per minute in the slow speed drives and 35 to 300 in the higher speed units.

Heavy porcelain insulation must be used in this type of flasher to prevent leakage to ground and consequent short-circuiting of the transformer secondary. The contacts may be used at any voltage up to 15,000 volts. The current capacity is limited to 30 milliamperes in most commercial units of this type.

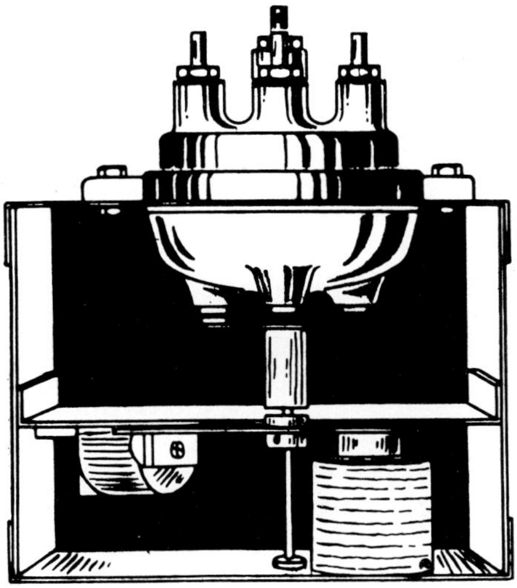

Figure 12.4. High-voltage or rotating-gap flasher. This type of flasher operates a number of luminous tube circuits from one transformer.

Since the rotating-gap type of flasher operates in the high voltage side of the transformer, special care must be taken to protect the transformer from severe changes in loading, that is, to avoid overload or underload as the sections of the flasher are connected in succession. The first rule is that the load should never be taken off the transformer secondary at any time by the switch. This means that none of the contacts may be left open, but all must be connected to a load, even if this means that two adjacent contacts must be connected to the same piece of tubing. Moreover, the flasher must be so constructed that the rotating element does not pass completely from under one stationary contact until it has passed under the next adjacent contact.

When designing a circuit to carry a high voltage flasher, it is necessary to take into consideration the maximum number of feet of tubing that will be employed in order to determine the transformer size. For example, if a high voltage flasher is connected to a series of arrow lines of tubing to give the effect of a contracting and expanding arrow and is operating at a high rate of speed, the transformer should be designed to carry the total number of feet of tubing in the arrow, even though at one sequence of flashing it may only carry one small section.

The use of high voltage or secondary circuit flashers results in a saving in tubing life. This is because the tubing is lit only for a comparatively short period of time compared with an installation where the tubing burns continuously.

When a high voltage flasher is used, some loss of the transformer voltage must be expected in breaking down and maintaining the spark between the rotating arm and the contact. It is estimated that the value of voltage required to break down this gap is approximately 1000 volts. This should be taken into consideration when designing the transformer load. The high tension sparks present in this type of animater produce ozone from the oxygen in the air. Ozone is active chemically and may combine with the nitrogen and moisture in the air to form dilute nitric acid, which may injure metal and rubber parts nearby. Protection of such exposed parts with a coating of asphalt paint will usually prevent this trouble.

Animation Diagrams. The types of animation possible with primary or secondary circuit flashers are virtually without limit. So many ingenious flashing sequences have been devised by sign manufacturers to suit the requirements of particular installations that it is impossible to describe even a small fraction of them.

There are several basic animation diagrams which are so widely used that they should be familiar to every sign manufacturer. These are on and off, alternate, simple speller action, and other standard flashing patterns. To determine the necessary sequence of a flasher for ordinary purposes, standard charts are used which show the number of contacts or sections desired and the flashing sequence for one complete cycle. Figure 12.5 shows a number of different charts of flashing sequences that can be employed for a speller-type flasher. Every manufacturer can use a similar chart to show the length of time, the number of contacts, and the flashing sequence for the type of flasher that has to be designed for this particular job. In Figure 12.5, Chart A indicates a speller flasher that has five contacts or sections. Section one goes on first, then 1½ seconds later section two goes on, then 1½ seconds later section three, then four, then five. As each section goes on, it continues burning. One and one-half seconds before the completion of the cycle all circuits go off, until the end of the cycle, when the flashing sequence is repeated. On the other hand, in Chart C, section one goes on for two seconds, but then goes off when section two goes on, and this goes off when section three goes on, etc., until the end of the cycle, when the spelling action is repeated.

Figures 12.6, 12.7, and 12.8 are large signs indicating various types of animation that can be obtained with flashers. By skillfully designing a sign to employ flashers of any of the types described, an almost endless variety of effects may be obtained.

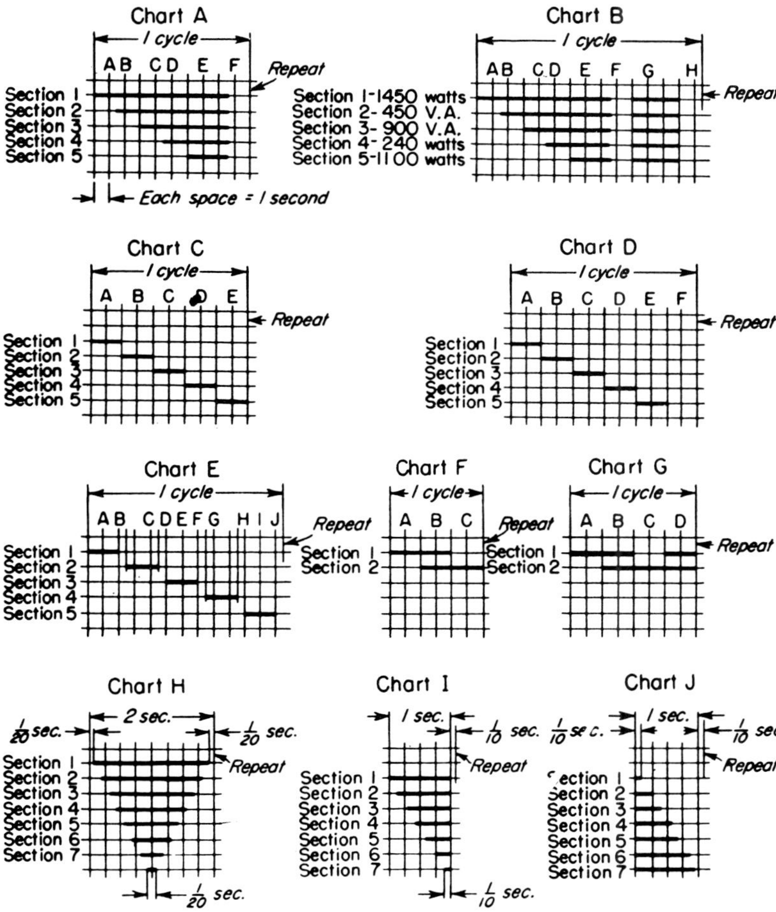

Figure 12.5. Typical sequence charts for speller-flasher operation as furnished by the Time-O-Matic Company as a guide.

Time Switches. An important type of primary switch is the time switch (Figure 12.9), which is used to turn signs on and off at predetermined times. Essentially, the time switch is an electric clock fitted with contacts which control the load, or which may

Figure 12.6. A spectacular weather-indicating sign designed by Douglas Leigh.

Figure 12.7. A neon sign combined with a jump clock for direct reading of time in hours and minutes.

Figure 12.8. A spectacular luminous tube sign. Two rows of tubing running vertically and alternating by rapid flasher action produce the effect of coffee pouring into the cup. Smoke from the cup is produced by a special smoke generator.

actuate an auxiliary contactor which in turn controls the primary circuit of the sign. The electric switch operates on a 24-hour cycle. A dial connected to the clock shaft is marked with two 12-hour scales, one representing day, the other night. The dial is fitted with two pointers, one of which is set opposite the time at which the sign is to be turned on, the other opposite the time of turnoff. The load contacts are capable of handling 15 to 25 amperes per pole, depending on their size and construction. Auxiliary contactors are available for any load up to 500 amperes per pole. Most time switches are arranged so that they may be operated manually at any time without interfering with the normal automatic operation. Special types of switches are available which automatically advance the turnoff time with the season, as

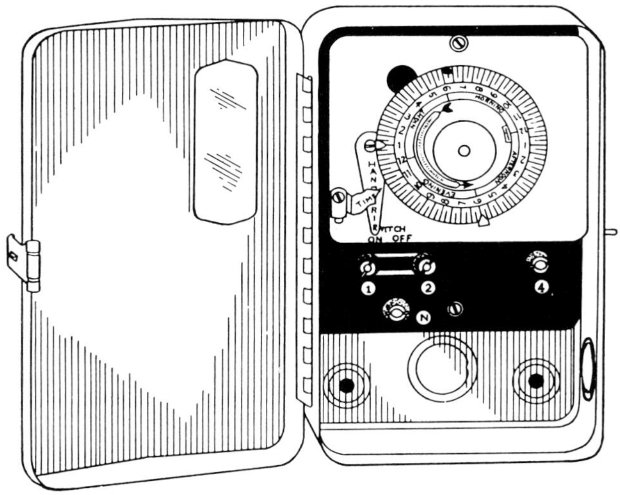

Figure 12.9. A time switch which can be set for turning a sign on and off at regular intervals.

the days grow longer. These are expensive units, however, and are not worthwhile unless the sign must operate entirely without attention for long periods of time.

The photoelectric switch uses a phototube (electric eye) to control the sign directly in terms of the amount of daylight present. Such switches automatically follow the change in the length of the day and, in addition, may be relied on to turn on the sign whenever the daylight fails, during storms, and on particularly dark, cloudy days.

CHAPTER XIII

Radio and Television Interference

In many locations neon signs or even interior lighting installations may cause interference in nearby radio sets and television receivers. The elimination of radio or television interference may be one of the most perplexing problems facing the neon signman. The reason for this is that interference from high voltage signs or lighting circuits may be due to many different factors. In order to eliminate interference, it is necessary to understand why the interference is caused and to take the proper steps to remedy it.

In general, interference may be caused by certain defects inherent in the sign itself or by its position in relation to the set. If the sign has flashing units, the interference may be directly due to the flasher mechanism. On the other hand, it may be due to a combination of factors, and it is for this reason that it may be necessary to follow a certain organized procedure to determine how the interference is caused.

The order in which the steps are listed is very important. It is based on taking the easiest and most effective measures first and then going on to the more difficult steps that may sometimes be necessary. Practical experience in the field shows that a great deal of time is usually saved by following the sequence recommended here.

The first step is to make sure that the sign is actually causing the interference. Second, try to move the aerial and line cord of the receiving set away from the sign. If this does not help, the third step is to ascertain whether or not a flasher is causing the trouble. If it is, proceed directly to step 17, which deals with flashers.

If the interference is not caused by flashers, go on to step 4 which will show which section of the sign is at fault. Step 5 is to check the grounding connection of the transformer. Step 6 shows whether the low voltage circuit leading to the transformer is faulty. The transformer itself is tested for defects in step 7. The next step is a test to determine whether the transformer is overloaded, and if so, why. In step 9 the high voltage circuit is checked for loose connections.

Step 10 checks the grounding of metal parts of the sign which should be grounded, while step 11 ensures that metal parts which should never be grounded are actually not in contact with grounded objects. In step 12 the spacing and insulation of adjacent pieces of glass tubing are inspected and corrected. We make sure that dirt is not causing the trouble in step 13, and the same is done for paint in step 14. Next, porcelain housings, if they are used, are examined for defects. Step 16 is to install a filter in case the sign is located in such an inaccessible position that it is impossible to get close enough to take any of the preceding steps.

Steps 17 to 20 deal with interference caused by flashers. After each step it is necessary to return to the receiver to observe whether the interference has stopped. As soon as it stops it is unnecessary to proceed any further. In most cases this will occur before half the steps have been taken.

Step 1. Whose Sign Is at Fault. Turn off the sign and see if the interference stops. Frequently much time is wasted before discovering that a sign in another locality is at fault.

Step 2. Moving the Aerial. The location of the aerial in relation to the sign may cause a greater pickup of interference. If the aerial is moved, or rotated to a different position, the interference may be eliminated or greatly reduced. In the case of roof signs interfering with television reception, the antenna may be aimed directly at the sign. It may be possible to change the orientation of the antenna so that it is not directed at the sign. In some cases the receiver plug-in cord may be acting as an antenna and picking up the interference because it is too close to the sign. This type of interference is encountered with skeleton signs or outline borders, where the neon sign installation may be quite close to the receiver.

Step 3. Determining the Type of Interference. There are two types of interference: interference caused by arcing and glow discharge and interference caused by flashers. The first shows up as a continuous buzz in radios and as continuous snow on television screens. The second type, flasher interference, causes distinct clicks synchronized with the flasher action in radios and as snow or distortion synchronized with the flasher in television. The first type may be located and remedied by steps 4 to 16, and the second type by steps 17 to 20. If both types of interference occur together, it is advisable to eliminate the flasher interference first.

Step 4. Picking Out the Bad Circuit. If a sign contains a single transformer, there is no difficulty in determining the circuit at fault. But if a sign contains several transformers, then it is necessary to locate the transformer circuit that is defective. In this case each transformer circuit must be turned on separately.

Step 5. Checking the Transformer Ground. When the interfering circuit has been found, it is necessary to make sure that the transformer has its mid-point securely grounded. Most transformers of 7,500 volts and over have the mid-point of the secondary grounded to the metal case. If the transformer is securely fastened to the metal sign, the transformer should be automatically grounded. However, some transformers may have a separate post brought out for grounding purposes, and this should be securely fastened to a good ground. Also, some transformers may be placed away from the sign on insulated material. In this case, a separate wire should be connected between the transformer case or terminal and a good ground.

Step 6. Checking the Low Voltage Circuit. Disconnect the primary leads from the transformer in the defective circuit and connect them to an incandescent lamp. If the interference continues, it means that the trouble is in the line leading to the transformer primary. This can be caused by a loose connection in a panel box, a loose fuse, a break or faulty splice in the line, or bad or wet insulation. If water has run into the conduit in which the line is located, interference can result. After this step, connect the line securely to the transformer terminals.

Step 7. Checking the Transformer. A defective transformer may cause interference. The transformer may be checked by operating it in both open-circuit condition and short-circuit condition. To operate on an open circuit, disconnect the cable leads at the high voltage terminals of the transformer and turn on the primary voltage. To short-circuit the transformer, connect a piece of wire across the high voltage terminals, and then turn on the primary voltage. If interference occurs in either open-circuit or short-circuit condition, it indicates that the transformer is defective and should be replaced.

Step 8. Checking for Overload. Short out one or two units of tubing carried by the transformer. If the interference stops, the transformer is overloaded. An overload can be caused in four ways. (1) The length of tubing may be too great. (2) The pressure in one or more units may be excessive, either because too much gas was originally inserted or because air has entered the tube. (3) The pressure may be too low because not enough rare gas was inserted when the tube was pumped or because the sign has been operating for many thousands of hours and the rare gas has been "cleaned up." (4) The length of high voltage cable in the circuit may be too long.

If there is too much tubing on a transformer, it is necessary either to distribute the load properly, or to install an additional transformer. Excessive pressure is sometimes difficult to detect. If it is high enough, it will cause the discharge column to become a thin line in the center of the tube instead of a diffused line occupying almost the entire tube width, as shown in Figure 13.1. This can be observed through a piece of clear tubing near an electrode or at a splice. In addition, a high pressure tube runs hotter than one at correct pressure. When air enters a mercury-filled tube, it produces a reddish glow in place of the blue mercury glow until it diffuses sufficiently or is cleaned up. In the case of neon tubes containing no mercury, air causes a bluish glow in place of the red neon glow. A tube containing several millimeters of air runs very hot. It is extremely important to detect a leaky unit because if it is left on the sign it will continue to leak and its resistance will continue to rise. The sign inevitably will cause more trouble and go out altogether, making it neces-

Figure 13.1. Diffuse discharge in top illustration shows correct pressure in tubing. Concentrated discharge in bottom illustration indicates a tube with high gas pressure.

sary to return and perform another servicing job. Moreover, if a high resistance unit remains on a sign for some time, it may cause the transformer insulation to fail.

Insufficient rare-gas pressure can be detected after the condition has existed for a few hundred hours by noticing whether the glass near the electrodes has blackened inside the tube. The blackening results from a deposit of metal sputtered off the electrode. Low pressure greatly accelerates electrode sputtering, and the sputtering itself, if it becomes severe, can cause radio interference. Electrodes with a ceramic collar on the shell, however, resist sputtering under even the worst conditions.

Too long a length of high voltage cable can cause interference. When long high-tension leads must be used, extra long cable supports should be employed to get the cable as far from grounded metal as possible. In cases where long returns are used, that is, over 20 feet of high-tension wire, or over 10 feet of high-tension wire in a metal shield, such as Greenfield or pipe, make the following test: Disconnect the long length of cable from the tubing and connect it to the transformer so that the cable forms a short circuit across the transformer. Then, if the interference continues, despite the fact that the tubing is no longer in circuit, and there was no interference in step 7 when the transformer was short-circuited by a short piece of cable, the trouble is obviously caused by the long leads. In that case, it may be advisable to divide the load and use two smaller transformers of lower voltage. With long high-tension returns, extra long cable supports should be employed to get the cable as far from grounded metal as possible. When cable must be placed in a sheath or conduit, every effort must be made to make the leads as short as possible.

Step 9. Inspection for Loose Connections. Inspect for loose connections in the high voltage circuit. Loose connections produce arcing and are caused by corrosion, vibration, or faulty installation. Corrosion occurs frequently in regions near the ocean or near factories and mills that give off corrosive smoke or fumes. Copper wire is very susceptible to corrosion. Sometimes a loose connection occurs inside a housing because the electrode does not make contact with the housing spring. This will cause an arc which will become worse in time.

Step 10. Grounding Metal Parts Securely. Make sure that all metal parts of the sign which do not touch luminous tubing or high tension cable, such as boxes, channels, or frames, are grounded solidly. If parts are grounded by screwing down a grounded wire, it is important that rust, dirt, or corrosion does not prevent the wire from making contact under the screw. Such grounding points should be soldered, as well as screwed tight, to ensure permanently good contact.

Nearby metal objects can also cause interference even though they are in no way connected to the sign. An electric charge can be induced in a nearby structural beam, rain gutter, metal venetian blinds, decorative metal strip, etc. Sometimes so much electricity is induced in neighboring objects that it is possible to get a shock by touching them. The only solution is to ground such objects or move them away.

Step 11. Keeping Tie Wire Away From Ground. Metal parts which are in contact with tubing or cable must be well insulated from ground. Tie wire used to fasten tubing and cable-to-cable supports must be secured to prevent it from getting near grounded metal. When window signs are hung from chains, the chain should not contact tubing or cable. When this rule has been violated and it is too difficult to move the chain, it may be possible to reduce the interference by insulating the chain from ground and from any other chain which touches the tubing or cable at a different point.

Step 12. Insulating Adjacent Pieces of Tubing. In bend backs at least half an inch of air space should be left between adjacent tubing. This is especially important with letters larger than a foot high and less important with smaller letters. When an improperly constructed sign has insufficient space at bend backs, the insertion of several thin sheets of mica will reduce interference. In case adjacent tubing is tied together for strength, as in Figure 13.2, fish line or string should be used instead of tie wire, since wire could cause arcing.

Step 13. Removing Insects and Dirt. An accumulation of dirt and insects inside housing can cause very severe interference. Also, clean the dirt off cable and tube supports, tubing and cables, since dirt greatly lowers surface resistance and promotes interference.

Step 14. Checking Paint. Painting housings with the same paint as the rest of the sign is a bad practice because it substantially reduces the surface resistance. On tubing only high quality block-out paint should be used because low resistance paint can

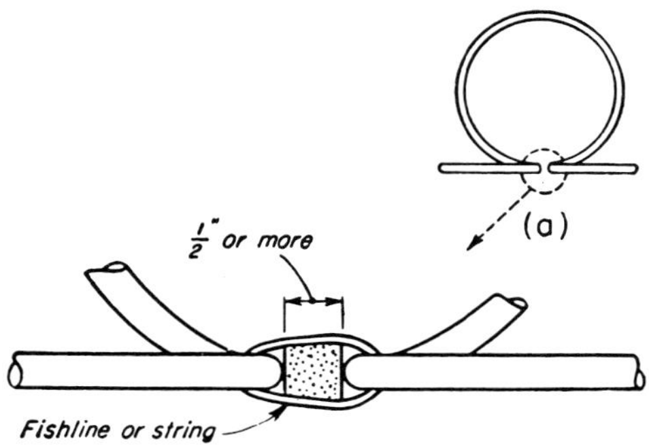

Figure 13.2. Tubes should be insulated from each other when connected together to strengthen the unit.

cause trouble. If it is possible to get a shock by touching the paint, it is a sure sign that the paint has low resistance, and better paint must be used.

Step 15. Checking Porcelain Housings. Old or defective porcelain housings sometimes cause trouble because water may seep through a small crack in the glazed surface and permeate the interior. This reduces the housing resistance so much that leakage currents are conducted through the porcelain and generate interference. Pyrex housings are not susceptible to this defect.

Step 16. Filter May Be Necessary. The measures recommended in steps 5 to 15 will remedy or suppress interference caused by inherent defects in the neon sign circuit, that is, interference due to arcing or glow discharge. However, in some cases the sign may be located in such a position that it is difficult to check according to the procedure recommended. Also, if the interference is mild, it may be suppressed by the use of filters which are placed in the circuit indicated in Figure 13.3. Severe interference cannot in all cases be completely suppressed by the use of filters alone, and it is necessary to find the source of trouble and remedy it at that point.

The following steps deal with interference caused by flashing or switching mechanisms.

Step 17. Flasher Must Be Enclosed in Metal Box. Some flash-

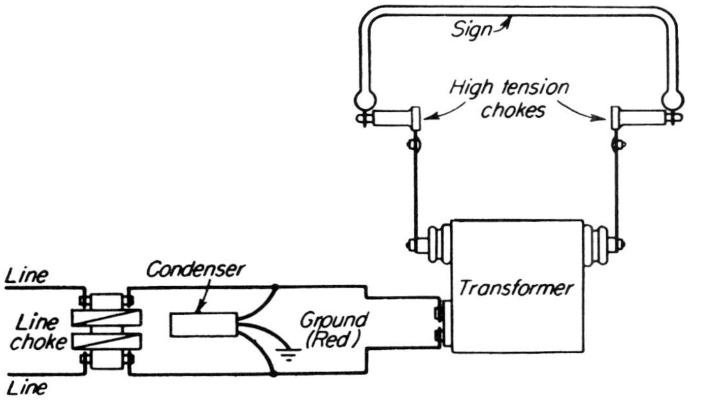

Figure 13.3. Filter circuit to eliminate interference caused by mild arcing or glow discharge (step 16).

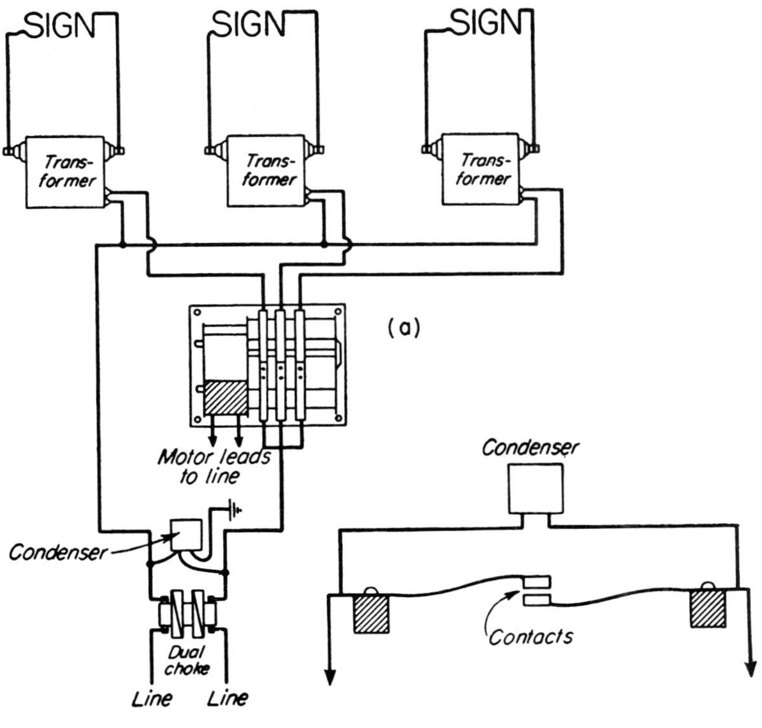

Figure 13.4. Filter circuit across the line as used with a flasher. A condenser is also connected across each set of flasher contacts.

ers are built in a metal box and are provided as a complete unit. Most of the high voltage flashers are supplied in this manner; however, many of the low voltage or primary flashers do not come in metal boxes. Every flasher, whether a primary or secondary unit, should be completely enclosed in a grounded metal box.

Step 18. Inspection of Flasher Contacts. Inspect the flasher contacts to see if they are severely burned or pitted. If so, they must be replaced. Make sure that the contacts on a low voltage flasher are closing tightly and making a good contact with each other. Poor contact on oxidized surfaces may cause arcing when the two contacts are supposedly closed. This may cause not only radio interference, but a flickering sign. High voltage rotary-flasher contacts are not desired to close fully, but should have a space of approximately 1/32 inch between them.

Step 19. Installation of Condensers. In the case of low voltage flashers, place a condenser across each set of flasher contacts. This tends to prevent the arc from radiating interference each time the contact opens. The condenser is connected across each set of contacts of the flasher as shown in Figure 13.4.

Step 20. Installation of a Filter Circuit. Even after all the steps described above have been taken to eliminate or correct flasher interference, it may be necessary to employ a filter circuit. The filter circuit consists of a combination of condensers and choke coils. A filter for a low voltage flasher is shown in Figure 13.4. A

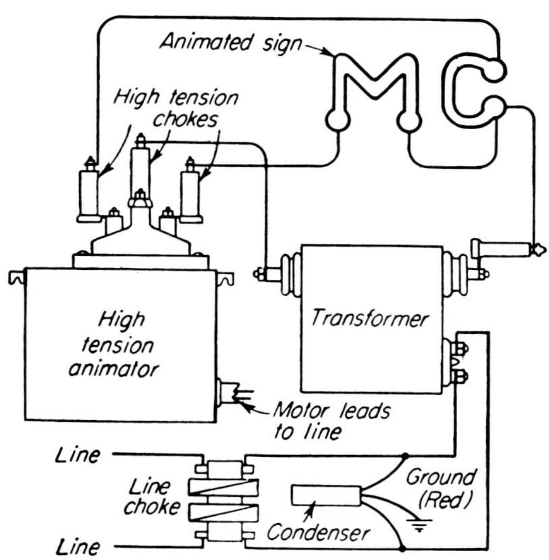

Figure 13.5. High voltage flasher interference may be eliminated by using chokes and condensers as shown.

filter for use in a high voltage flasher circuit is shown in Figure 13.5.

The measures recommended in this chapter are thorough enough to eliminate or suppress interference in almost any situation. After a certain amount of practice with the procedure outlined, it will be found that the time required to stop interference is much less than that needed when haphazard methods are used. It is well to bear in mind the contents of this chapter while in the process of constructing the sign itself, so that causes of trouble can be avoided in advance.

CHAPTER XIV

Installation and Maintenance

Sign Installation. The problem of installing the completed sign at the desired location can rarely be solved by following a fixed set of rules. Each type of sign and each location offers its own particular problem. Sign hanging is a dangerous business unless carefully done, and in busy districts the hazard to the public is considerable unless the work is done by competent men.

It is not the purpose of this chapter to provide all the information necessary to hang or to install signs. But the information given will be of considerable help to installers.

Safe Working Loads for Sign Hardware. To promote the best possible use of sign hanging material, Table 14.1 has been prepared indicating the safe working loads for various items used in hanging signs. The values given include a safety factor which takes into consideration recent engineering standards. It is important to emphasize that the safe working load depends not only on the hardware used but also on good workmanship.

Sign Maintenance. The service problem in the neon industry has never received the attention that it deserves. It has been assumed that the major problem was that of producing and installing the signs and that, if they were properly made, the maintenance problem would take care of itself. If the reputation of the sign manufacturer with his customers is to be preserved, proper reserves both of equipment and of capital must be made to take care of servicing signs. Accidents will happen to even perfectly made signs. When they do, the manufacturer must be ready to repair the damage quickly and well. Service is paramount, since a defective sign has lost its advertising value.

The problem of maintenance can be attacked from two points of view. The first is to produce and install the signs with an eye to the future so that the calls for service will be reduced. The second is to familiarize oneself thoroughly with the major causes of failure so that when a sign breaks down the nature of the trouble can be analyzed at once and corrected with the least delay and expense. The sign manufacturer should keep both of these points of view in mind.

Causes of Service Calls: Tubing Failures. There are two places where a sign may fail: in the transformer and its connections, including switches and fuses, and in the tube itself. Tubing failures may arise from breakage of the glass, from sputtering and cleanup of gas due to improper gas pressure, from

Table 14-1. Safe Working Loads for Sign Hardware

(Sizes in inches, loads in pounds)

To promote the best possible use of sign-hanging material, the following tables can be used for safe working loads. The values given include a safety factor which is made with recent-date engineering standards. All safe working loads depend not only on the item itself, but also on the other materials used with the article, and also on careful workmanship.

Stranded Cable

Diameter	*Safe Load*
$\frac{3}{16}$	1,150
$\frac{1}{4}$	1,900
$\frac{5}{16}$	3,200

Sign Chain

Size	*Safe Load*
3-0	1,280
4-0	1,500

Sign-hanging Bolts
(Iron bolts with square nuts)

Size	*Safe Load per Bolt*
$\frac{3}{8} \times 1$	820
$\frac{3}{8} \times 1\frac{1}{2}$	820
$\frac{1}{2} \times 1$	1,520
$\frac{1}{2} \times 1\frac{1}{2}$	1,520

Wire Rope Clips

Size	*Safe Load*
$\frac{3}{16}$	179
$\frac{1}{4}$	511
$\frac{3}{8}$	1,306

Nut-eye Bolts

Size	*Safe Load*
$\frac{3}{8} \times 2\frac{1}{2}$	820
$\frac{1}{2} \times 3$	1,520

Turnbuckles

Size	*Safe Load*
$\frac{1}{4} \times 4$	312
$\frac{5}{16} \times 4\frac{1}{4}$	540
$\frac{3}{8} \times 6$	820
$\frac{1}{2} \times 6$	1,510

Lag Bolts

Size	*Safe Load*
$\frac{3}{8} \times 2\frac{1}{2}$	820
$\frac{1}{2} \times 3$	1,520
$\frac{1}{2} \times 4$	1,520

*Wall Hooks**

Size	*Safe Load*
4	100
6	150

Screw-pin Anchor Shackles

Size	*Safe Load*
$\frac{1}{4}$	511
$\frac{5}{16}$	911
$\frac{3}{8}$	1,306

S Hooks

Size	*Safe Load*
$\frac{1}{4} \times 2$	160

Expansion Shields†

Long standard sizes		Short standard sizes	
Size	Safe load	Size	Safe load
$\frac{3}{8} \times 2\frac{3}{4}$	225	$\frac{3}{8} \times 2$	175
$\frac{1}{2} \times 3\frac{1}{2}$	375	$\frac{1}{2} \times 2$	325
		$\frac{5}{8} \times 2$	400

Table 14-1. Safe Working Loads for Sign Hardware (Continued)

Stove Bolts

Size	Safe load	Size	Safe load	Size	Safe load
3/16 × 1/2	112	3/16 × 1 1/4	112	1/4 × 1	320
3/16 × 5/8	112	3/16 × 1 1/2	112	1/4 × 1 1/4	320
3/16 × 3/4	112	1/4 × 1/2	320	1/4 × 1 1/2	320
3/16 × 7/8	112	1/4 × 5/8	320	3/8 × 1	820
3/16 × 1	112	1/4 × 3/4	320	3/8 × 1 1/2	820
		1/4 × 7/8	320		

* The same considerations apply to wall hooks as expansion shields, and wall hooks must be located at least 4 inches apart.

† The controlling factor in safe loading of expansion shields is the strength of the brick or concrete. For this reason, holes for expansion shields should not be located closer than within 6 inches of each other. Otherwise a large load might be concentrated on a small region of brick or concrete and cause failure.

leaks either at the electrode seals or elsewhere in the tube, from broken connections between tubes, and from blackening of the glass (especially in fluorescent mercury tubes). All these causes of trouble can be found by simple inspection or else by the application of a testing coil. They can be prevented to a large extent by the proper manufacture and installation of the tubing. Throughout the previous chapters, precautions against failure have been emphasized, and if the directions given are followed, tubing should stand up for its full life.

At the end of the useful life of the tubing, it must of course be replaced. But before the tubing unit is taken out and repumped, the precaution should be taken to determine whether the tube is really going bad or if poor connections or other electrical causes are at the root of the trouble. It is wise to take a spark coil to the job, fitted with a long extension cord. The coil can be used to test the tubing while it is still in the sign. If the tubing unit shows no sign of life when the sparker is applied to it, then that unit is defective and must be replaced. If it glows and shows signs of being normal, then the trouble must be sought elsewhere. A defective tube may be caused by poor electrical connections at the electrodes. This cause of tube failure should be rectified before the repaired tubing is replaced.

Electrical Failures. The second class of trouble lies in the transformer and its connections. In a recent survey of sign transformer troubles, made by one of the largest manufacturers of this type of equipment, it was found that 95 per cent of all the trouble with transformers could be traced to four causes: insufficient protection from the weather; transformer operated at incorrect frequency, as when operated from a converter running below normal speed or with excessive output voltage; improper transformer loading; and operation on too high a primary voltage.

To determine whether or not the transformer is in working order, an alternating-current milliammeter (see Figure 14.1) may be connected directly across the secondary terminals. If the rated short-circuit current flows, the transformer is in good condition. If no current or less than the rated short-circuit current flows, the transformer is out of order and should be replaced.

Occasionally when a high voltage flasher is used in series with the tube, this device will cause considerable trouble. The trouble can be found almost immediately by short-circuiting the flasher terminals. If the tube glows to the proper brilliance, the switch is at fault, and it should be repaired or replaced. Because of the many different varieties of flashers which are used, it is difficult to give any general rules for testing them, other than the above. When the cause of the sign failure has been determined, the defective part must be removed, repaired if possible, and replaced.

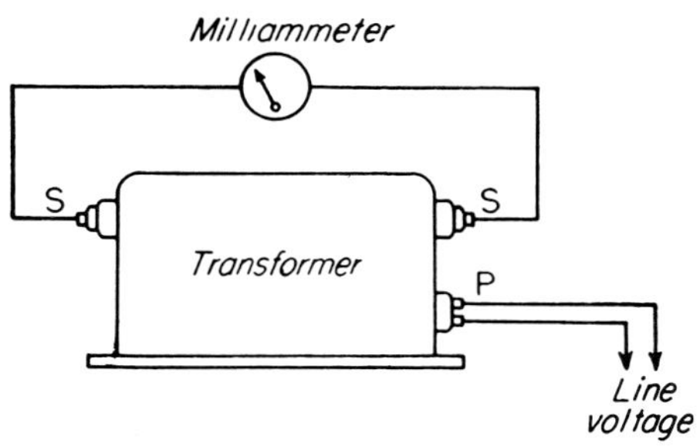

Figure 14.1. Test to determine if transformer produces rated secondary current.

Repairs and Replacements. Tubing. When tubing is at fault, the tubing section which has failed must be removed, as a whole, and replaced. But if the sign has already reached its life limit, and especially if the still-operating tube sections show signs of age, it is often wise to take down the entire tubing and replace it. Tubing sections must eventually fail, and it will often save money to replace them all at once rather than one at a time. Special precautions must be taken in repairing fluorescent tubing, to ensure a good color match and equal brilliance in the old and new sections. This subject is treated at length in the chapter on cold cathode fluorescent tubing.

The tubing section can be removed from a box-type sign when electrode housings are used, simply by removing the tie wires which fasten it to the elevation posts and then pulling the tube out of the housing. If electrode bushings are used, the connector between lead wires and the cable section must be disconnected.

Once removed, the tubing section is taken to the shop. If the glass is not broken, the tubing may be used again. In some cases, it may be advisable to replace the tubing, but usually the old tubing is still suitable for use. The first step is to remove the electrodes, which should not be used again.

Removing the electrodes is accomplished simply by breaking the glass with a file cut in the usual manner, at the joint between electrode jacket and the tube proper. It should be remembered when making this break that the new electrode, when spliced to the tube, must fit accurately in the position occupied by the original electrode before removal and also have the same height. This necessitates care in leaving neither too little nor too much glass on the tube where the splice will be made. The outside of the glass should be wiped clean to prevent the possibility of dirt collecting inside the tube.

Resealing Electrodes. New electrodes of the type used in the sign originally should be used. One electrode is spliced on in the usual manner. In this work, the original layout sheet, used for the glasswork when the sign was first made, should be used to ensure the correct position of the electrode with respect to the rest of the tube. When the first electrode is on, a new tubulation glass must be spliced on in the usual manner, and finally the other electrode is sealed on.

Special electrodes can be obtained for repumping all mercury tubes. These electrodes contain an extra stud wire coated with a chemical that releases a quantity of hydrogen during the bombarding process, which attacks the dirty mercury or mercury spots and prevents mercury stains. The intense heat will cause the hydrogen to combine with the impurities and the two are pulled out by the pumping system, leaving clean mercury in the tubing. Except for a decrease in light output due to age, the tube should operate like new.

Repumping, Bombarding, etc. From this point on, the pumping, bombarding, and filling of the tube are exactly the same as they were when the tube was first made. Although the job is a repair, it should be done with the same thoroughness and care used on the original tube. If the tube has failed before its expected life length was reached, special precautions should be taken to see that it is repumped carefully, since the original procedure may have been faulty. To prevent a recurrence of the failure, extra precautions should be taken to make the tube in the approved manner.

Tube Breakage. Tubes often crack under the influence of the high heat which follows gas cleanup. Also, they may break because of accidents to the tube, while it is in the sign. In these cases, it may be necessary to bend an entirely new tube from fresh glass, although, when the glass is broken between letters or near the electrodes, it may be necessary to replace only part of the glass. In this latter case, all good parts of the tube are spliced onto the new tube in the usual manner, the broken parts being replaced with new tubing.

When new glass is used, it is necessary to match that of the original tube to the new glass. Usually this is not a difficult procedure, but occasionally the glass in the original tube itself has a somewhat different texture from the new glass to be spliced onto it. In these cases, the entire tubing section should be bent anew. Different diameters of tubing in the same tubing section are very conspicuous, although they may not be noticeable if the difference in diameter is small. The glass diameter of the tubing in the sign should always be matched as closely as possible; otherwise the tubing sections remaining in the sign will not blend well with the repaired section.

Matching glass is often one of the most difficult problems in repairing tube sections, especially if colored or fluorescent glass is used. The properties of the original glass may have become changed from being exposed to heat and to the weather. If the new glass does not fuse readily to it, the splice will be weak and subject to breakage.

Experience is the best guide in deciding whether to repair a tube section in whole or in part. Money can be saved by following whichever of these two possibilities consumes the lesser amount of material and labor and presents the lesser possibility of subsequent failure.

Electrical Repairs. Occasionally the wiring of a sign goes bad, particularly if the high voltage cable has been exposed to the weather. In this case, it should be replaced throughout with new cable. But if the transformer or flasher is the cause of the trouble, the repair usually cannot be made except by the manufacturer of the equipment. In the case of transformers, the job is too expensive to be attempted even by the manufacturer. When a transformer or other piece of electrical equipment goes bad, the only remedy is replacement with a new unit.

Every progressive sign shop should carry on hand a quantity of different sizes and types of transformers capable of duplicating any transformer installed in any of the signs manufactured by the shop. A burned-out transformer should eventually be replaced by one of the same rating, provided, of course, that the original transformer was properly chosen in the first place. Often it is wise in replacing transformers to apply the loading tests ad flicker voltage tests outlined on page 204 to make sure whether or not the new transformer is properly loaded.

Flasher Failures. When the flashing devices of a sign fail, as can be tested by short-circuiting their terminals, there is usually no remedy but the installation of a new flasher. If the flasher is a simple on-off type, its replacement may be quickly made. But if it is of a more complicated variety, involving many different effects, repairs are difficult and usually may have to be made at the factory. While the flasher is being repaired, or while a new one is being ordered, everything should be done to place the sign in service at least temporarily until the new part arrives. The best method is to short-circuit the switch in such a way that the whole sign will be lighted continuously as an emergency measure until a new flasher can be procured or the old one repaired. When this step is taken, the owner of the sign should be consulted, and the possible increase in operating cost explained to him. Usually he will want his sign on, even though it will cost him more money for the short period.

When a time switch goes bad, a conventional hand-operated off-on switch may be substituted and proper explanations and instructions made to the sign owner so that he can arrange to have the sign turned on and off manually.

The Importance of Continuous Service. In the above paragraphs, great stress has been laid upon keeping the sign operating, even by temporary measures, when interruption of service occurs. Many neon manufacturers find that they can keep their business and grow steadily without resorting to such extreme measures of service. But it is certain that customer satisfaction and the public reaction to the sign are greatly impaired when the sign fails and remains dark any appreciable length of time. Usually the customer will appreciate the effort that the neon man makes in keeping the sign lighted in the face of difficulties, and this appreciation will lead to increased business.

Neon signs are attractive only as long as they are fully lighted. A sign with several sections dead or, worse, with one or more sections weak and flickering is worse than no sign at all. The public reaction to such a display is always negative, and the sign, far from being useful for advertising purposes, actually does harm. It should be left completely dark, although a double-face sign, connected on the same transformer, may have one piece of tubing dead on one side only, while the other side still has advertising value.

In the interest of the neon business, each neon manufacturer should make this fact clear to each of his customers when he first installs his sign. He should request the sign owner to report at once any damage, flickering, or dead sections in the sign. The neon man stands to lose nothing by this procedure. Since he will have to attend to a damaged sign sooner or later, he might as well do it before any great amount of harm has been done. If the damage occurs before the guarantee is up, the service is costly to the manufacturer. But, on the other hand, if the damage occurs after the guarantee period, the result is business for him. All in all, it is a wise policy to inform the buyer that a broken sign may do harm and advise him to have any trouble corrected at once.

Maintenance Guarantees. The average luminous tube sign, regardless of the kind of gas used or the complexity of the sign, is guaranteed for a period of time, as a rule, for one year. In extreme cases of competition, this limit is sometimes extended to two and even three years. Experience has proved that many signs will give three years of life and more, but the one-year guarantee may be the only sensible arrangement to make. The guarantee usually takes the form of complete service during the guarantee period, which includes replacement of all weak, flickering, or dead tubes and of any defective electrical equipment. It may also include repainting of the sign box and crossovers if either begins to shed the paint. In other words, the sign manufacturer undertakes to keep the sign operating as well and looking as well for a period of one year as when it was first installed. After the guarantee period is up, repairs are charged on a regular basis, which includes the cost of the materials and labor and a reasonable profit.

The Service Contract. Another form of service arrangement is finding considerable favor. It is the service contract. Under this agreement, service contracts are sold either to a man whose guarantee has expired or to a man who buys a new sign which is guaranteed for a year, at the end of which the service begins. In exchange for the service, he agrees to pay a contract price of so much per month for complete service to the sign, whether or not any repairs are necessary. This contract basis may continue for as long as the buyer wants the service, but since the sign will begin to wear out rapidly after two or three years of service, the contract rate increases as the sign grows older. In other words, in the service contract, the sign manufacturer agrees to keep the sign in constant repair at a definite sum per year, payable monthly. This amount depends on the size of the sign, the number of feet of tubing, the age of the sign, whether or not it is to be painted, and also whether or not it has a rotary converter.

The service contract is very desirable. After a sign manufacturer has been in business for some time, he may accumulate a great many service contracts which bring in a steady monthly income. By constructing the original sign so that it is sturdy and the tubing and materials of highest quality, actual servicing may be cut down to a minimum. Then the service contract is a very profitable item.

Since the contract rate calls for complete service, whether it be a simple tube section or the replacement of a whole battery of transformers, the manufacturer usually finds that it is best to install a well-made sign. Then, if it stands up, he gets the contract rate but has no material cost or labor cost for repairs. In this way, the contract repair rate puts a premium on good workmanship in the first place. Although the man who buys the sign may feel disgruntled at paying out a monthly rate for service which his sign may not need, he has always the comfort of knowing that if trouble does arise it will be taken care of without further cost.

Estimating Repair Costs. Before the retail price can be put on a complete sign, some estimate of the probable repair costs must be made, so that the extra amount can be added to the cost or the contract repair rate established. It is not the function of this book to establish business or accounting methods of figuring plant and material costs or to prescribe a fair rate of profit on the owner's investment. But the problem should be dealt with by each manufacturer.

Where competition is heavy, the price situation will take care

Job No.____________________
Type: double-face swing sign
Guarantee: Feb. 3, 1934 to Feb. 2, 1935

Name____________________
Address____________________
Telephone No.____________________

(north side)

(south side)

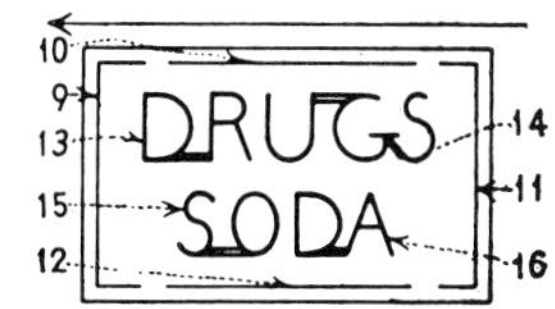

Record of repairs

Date	Defect	Nature of defect	Cause of defect	Cost of repair		
				Labor	Material	Total
2/10/34	Fuse		Small fuse			
6/7/34	No. 14	Cracked electrode	Defective housing			
10/20/34	Sign flickering	Loose connection	Poor soldering job			
1/25/35	No. 11	Broken	Broken by window cleaner's ladder			

Numbers in Defect column refer to glass tubing units indicated in drawings above.

Total cost of labor and material during guarantee period

Figure 14.2. Maintenance record card.

of itself; it is simply necessary to meet the quality and price of competitors and if possible to surpass their workmanship. But a fair price for sign manufacture, installation, and repair should be carefully arrived at by every progressive sign manufacturer. As an aid to this work, a complete record of equipment, materials, and labor cost should be kept, an inventory of stock taken regularly, and the cost attributable to each job estimated against the actual amount paid for materials, equipment, and labor. The use of the original layout in this work has been described on page 86. In Figure 14.2 is shown a maintenance card," which is also useful in handling repair estimates.

PART III

Fluorescent Lighting

CHAPTER XV

Fundmentals and Production Of Fluorescent Tubes for Lighting

The importance of fluorescent lighting, for general illumination purposes, is so widely appreciated that additional emphasis is hardly needed. Fluorescent forms of light have within recent years assumed a commanding position. The fluorescent lamp competes directly with the incandescent forms of light in the field of general illumination, offering a vastly more flexible light at considerably lower upkeep cost which, after a year's operation, makes up for the slightly higher cost of installation.

There are two forms of fluorescent lighting used: low voltage or hot cathode fluorescent lamps, and cold cathode fluorescent tubing.

The hot cathode form of fluorescent lamp is not manufactured by the signman or luminous tube specialist. The advanced techniques of manufacturing used by the larger companies allow them to produce the low voltage fluorescent lamp at prices which would be impossible to meet with the technique of a small luminous tube shop. However, the cold cathode fluorescent tube for lighting is a logical extension of the luminous tube practices of the tube bender. Its application as a source of interior lighting opens up a virtually new field to the sign specialist.

The Principle of Fluorescent Light. This chapter is devoted to the design, processing, installation, and maintenance of fluorescent tubes, particularly cold cathode fluorescent tubes. Before going into detail on these subjects, it is desirable to discuss the basic principles of fluorescent light. These basic principles apply equally well to the hot cathode and cold cathode types of tube.

Fluorescent light is produced by means of a coating on the inside of a luminous tube. The coating, known technically as a phosphor, is a mixture of chemical compounds which have the property of emitting visible light when subjected to rays of ultraviolet light. The phosphor is usually produced by combining one of the light metals such as zinc, calcium, magnesium, or cadmium with a compound such as silica or tungstic acid. Typical fluorescent powders are zinc, orthosilicate, calcium tungstate, magnesium tungstate, calcium silicate, and zinc beryllium silicate. Many other compounds and blends of several powders have been developed. Recently the hot cathode fluorescent lamp manufacturers have substituted halophosphates, essentially calcium-fluoro-chloro-phosphates, for the zinc beryllium silicates that were once used for white fluorescent tubes. To eliminate the beryllium from cold cathode and neon sign manufacture, calcium silicates and calcium alumino-silicates have been substituted for the zinc beryllium silicates in the production of white tubes.

A fluorescent compound must be carefully prepared if the light is to have a smooth even color, reproducible from lot to lot. Very minute amounts of certain foreign substances, such as lead or manganese, are used to activate the phosphor. The smallest change in these activating compounds will change the color. Even mechanical changes in the milling of the powder or the manner of application to the glass tube have a marked effect on the color, brilliance, and stability of the fluorescent light. For these reasons, the manufacture of fluorescent powders is a specialized business, based on carefully guarded formulas and procedures.

One obvious advantage of fluorescent light is the wide range of colors which can be obtained by employing different blends of powder. To this wide variety, further choice is added by the use of colored glass and by employing different gases as the exciting agents. With this wide range to choose from, it is no exaggeration to state that all colors possible on the artist's palette can be reproduced with luminous tubing. Many shades of white can be produced from commercially available powders, including one shade which is indistinguishable from daylight.

Important as this color flexibility is, it is outweighed, for interior illumination purposes at least, by the fact that the fluorescent tube is the most efficient source of light commercially available. The incandescent lamp, in the usual commercial wattage sizes, operates at a luminous efficiency of roughly eight per cent, whereas the fluorescent tube operates at nearly double this efficiency, above 15 per cent. This not only increases the amount of light produced per dollar for electricity, but it reduces proportionately the amount of heat generated by the lamp. The three advantages, wide color flexibility, more light per dollar, and less heat per lamp, explain the popularity of fluorescent lighting for general illumination, and they are equally important for display purposes.

The high efficiency of the fluorescent lamp is explained by the fact that the phosphor efficiently converts the invisible radiation present in the gas discharge into visible light. The gaseous discharge, whether it is neon or mercury in combination with argon, produces only a small amount of light visible to the eye. By far the larger percentage of the energy fed to the lamp is converted into invisible ultraviolet radiation. This is particularly true of the pure argon discharge which emits such a small amount of visible light that it is used only in conjunction with other gases or vapors. The mercury discharge is particularly rich in ultraviolet light.

The ultraviolet light serves no visual purpose because it is invisible. Most of this energy is ultimately converted into heat (see Figure 15.1). When the inner surface of the luminous tube is coated with a fluorescent powder, however, a considerable portion of the ultraviolet light is absorbed by the powder, which stores the energy for an extremely brief time (less than a millionth of a second) and then reradiates the energy at longer wave lengths which lie within the visible spectrum (see Figure 15.2). Thus the invisible ultraviolet light is converted into useful visible light.

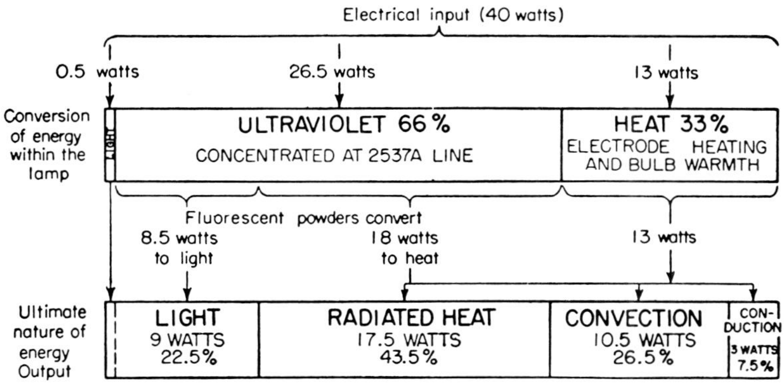

Figure 15.1. A 40-watt fluorescent lamp will develop 66 per cent ultraviolet light in the tubes. Only 22.5 per cent is converted into visible light and the balance into heat. (General Electric Company.)

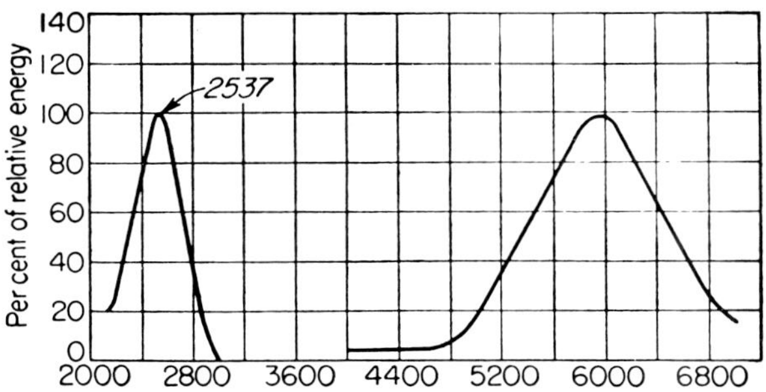

Figure 15.2. The curve on the left is spectral sensitivity with its peak at 2,537 angstroms. The curve at the right is the visual spectrum produced by ultraviolet light. (General Electric Company.)

The more efficient the phosphor, the more energy is converted within the visible spectrum. A white light phosphor produces light more or less uniformly over the spectrum, whereas a colored phosphor radiates light most strongly in a particular region of the spectrum. Unfortunately, since the fluorescent process is a delicate one, great precautions must be taken in the manufacture of the fluorescent powder and in its application to the tube, if it is to maintain its efficiency and its color over long periods of operation. With reasonable care in manufacture and processing, a phosphor can be depended upon to maintain about 70 per cent of its initial brilliance after 5000 or more hours of operation.

While the fluorescent tube is intrinsically a high efficiency device, compared with other forms of lighting, this advantage may not be fully realized unless the fluorescent lamp is mounted in a proper fixture, designed to conserve all the light emitted by the lamp and to direct it to the space where it will serve a useful purpose. For this reason, the reader is urged to consider carefully the principles underlying illumination design which are discussed in Chapter 16.

Characteristics of Fluorescent Powders. The primary characteristic of a fluorescent powder is the color it produces when irradiated by the ultraviolet light of the gaseous discharge. The phosphors of fluorescent powders can be grouped, as far as color is concerned, into two classes: basic powders and preblends. The basic powders are those which, by themselves, will give a distinctive color. In this group can be classed the green phosphor, zinc orthosilicate, which gives a brilliant green. This color is generally used for sign work only and rarely for lighting except for color effects. It has a luminous intensity about 30 per cent greater than that available from any other fluorescent material.

Calcium tungstate powder gives a distinctive blue, and the shade of blue can be slightly varied by the amount of the lead activator used. Magnesium tungstate will produce white and has been used mixed with other powders for white in hot cathode lamps. Calcium silicate produces a pink to reddish color and is now used in conjunction with other basic powders to produce white tubes. It has been developed to a very high degree in order to eliminate the zinc beryllium silicates also used as basic powders.[1] Cadmium borate gives a reddish color and is primarily used to produce red tubes of hot cathode type only.

The preblends consist of basic phosphors mixed together in the proper proportions to produce various colors. Almost any kind of color could be produced by blending the basic powders. It has been necessary, however, to limit the colors to a range that would be standard for practical purposes. Every manufacturer has tried to produce a similar line of colors. This has been

[1]The use of this zinc beryllium silicate basic powder is being gradually eliminated for all types of fluorescent tubes.

especially true in the range of whites. The standards for fluorescent lighting are generally limited to 3500 white, 4500 white, 6500 white or Daylight, and Soft White. Many other white tubes are available that can be used for special interior effects or for display or sign work, but for all practical purposes, the four whites indicated above are almost universally used.

Although not generally employed, colored glass can be coated with basic phosphors to give a soft or subdued lighting effect. For example, noviol glass, coated with one of the shades of zinc beryllium silicate or with calcium silicate, can be used to give a subdued gold effect. Dark blue or soft red colored glass can also be coated with these powders to give either dark blue or dark red subdued lighting effects.

Other basic fluorescent powders which are used primarily for fluorescent lamps are the halophosphates. They have approximately the same characteristic light output as the zinc beryllium silicates, emitting in the yellow-white and orange-white portions of the spectrum. Sign tubing coated with these chemicals may darken very badly at the bends because the phosphor may become damaged at the bent portion of the glass. Also extreme care must be taken when processing this type of coated tubing by internal bombarding because extreme heat may damage this type of phosphor. These powders are used extensively in mass-produced hot cathode lamps where tubing is never bent and is oven baked, rather than bombarded or heated internally. Experiments are being continually conducted concerning the use of halophosphates in cold cathode tubing because of the high lumen output of this class of powders. It is possible that the difficulties encountered in bending and processing by internal bombardment may be eventually overcome.

In the past there has been a tendency to use "black light" for illuminating both interior decorative walls and outdoor bulletin boards, which are painted with fluorescent paints. Fluorescent paints, which respond to the black light or ultraviolet radiation, consist chiefly of sulfides, of which the most important are zinc sulfide and zinc cadmium sulfide, activated with copper. The paints are excited most effectively by radiation of 3650 A. This radiation is abundantly produced by mercury-filled tubes coated with the 360 BL phosphor, cerium-activated calcium phosphate. However, visible light as well as ultraviolet is radiated by this phosphor, and for the many applications where visible light is undesirable, it is necessary to use either a filter or tubing that passes only ultraviolet light. Furthermore, ordinary lead-glass tubing cannot be used for "black light" tubes because it does not transmit ultraviolet light. Black Light Blue glass tubing, manufactured by the Corning Glass Company, is well suited for these lamps since it passes ultraviolet light, does not pass visible light, and seals easily to lead-glass electrodes. This glass also can be bent like lead glass.

Tricalcium phosphate, barium aluminate, and barium alumin silicate emit light in the ultraviolet and near-visible portions of the spectrum. Good units can be made of nickel cobalt or light blue tubing coated with any of these powders. Units made of clear ultraviolet transmitting glass in conjunction with blue filters are also employed.

Zinc aluminate coated on ultraviolet transmitting glass produces radiation in the erythemal region used for "health" and "sun" lamps.

Phosphorescent sulfides absorb both visible and ultraviolet radiation and later emit visible light when the activating source is removed. The afterglow of some of these substances can be seen in darkness for some time. Zinc sulfide has a high initial phosphorescence but loses its glow after an hour or so. Calcium and strontium sulfide have a weaker initial phosphorescence but can be seen in total darkness as long as 8 to 12 hours.

Although not in general use, the various "black light" or ultraviolet powders, briefly described above, may become a big factor eventually in lighting displays (see Figure 15.3).

Luminous Efficiency of Phosphors. The second important characteristic of phosphors is their luminous efficiency. In general, the amount of light produced by a fluorescent lamp increases with the current through the lamp. In the attempt to obtain maximum brilliancy, some cold cathode manufacturers may use high current transformers with ratings higher than 120 milliamperes. Unless the proper size tubing is used, the light does not increase indefinitely as the current is increased. The prob-

15.3. A bulletin board painted with special paints and using black-light lamps and spotlights. The black-light lamps are lit all the time, while the spotlights flash on and off, producing distinct changes in appearance and wording.

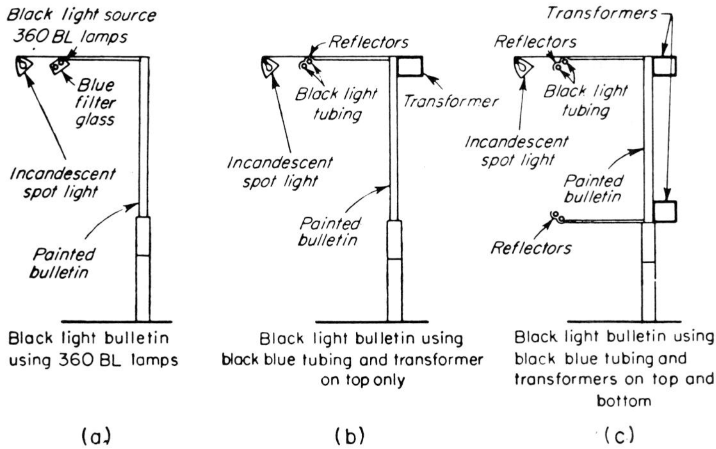

Figure 15.3A Cross sections of a bulletin. (a) Using hot cathode lamp units. (b) Using cold cathode tubing on top only. (c) Using cold cathode tubing on top and bottom.

lems, especially of electrode heating and disintegration, which accompany high currents must be considered in great detail.

When the color characteristic and luminous efficiency of a phosphor have been specified, it might be thought that the suitability of the material could be judged without further information. But this is far from the truth. A very important attribute of any fluorescent powder is a hard-to-measure property known as *stability*. A stable fluorescent material will maintain its light output nearly constant after several hundred hours of operation, whereas an unstable compound will lose brilliance rapidly. The stability of a phosphor is difficult to measure directly since the gradual loss of light may be hard to measure by the eye. If a photographic exposure meter is set up at a given distance from a section of tubing, the change in brilliance in the first hundred hours may be readily noted. When a section of a fluorescent tube is repaired with a new section, the contrast is especially noticeable, so much so in fact that it may be necessary to replace other sections in order to avoid this effect.

The stability of a phosphor is tied up with its chemical composition, the method of manufacture, and the manner of application to the glass tube. It depends on the manufacturers who compound the powders and those who coat the glass tube with them. Reputable manufacturers can be depended on for stable materials. The stability may be injured by careless processing, especially by rough handling in bending, and by carelessness in pumping and filling the tube.

Many phosphors, particularly those offered to the trade in the early days, showed a 50 per cent loss in brilliance in the first 200 operating hours. As manufacture has improved, this loss in brilliance has been reduced to about 25 per cent in the first 500 hours, with occasional losses running as low as 20 per cent in 1000 hours. All phosphors display this loss in brilliance to some extent. It is important to minimize the effect by the purchase of the highest quality materials and by employing extreme care in handling the coating throughout the processing of the tube. It is usually better to choose a stable phosphor, even though its initial brilliance may not be so high as that of a less stable competing material.

Manufacture of Fluorescent Coatings. In Chapter 3 the various methods used to coat the interior of glass tubing with a fluorescent powder were described. When using fluorescent tubing for cold cathode lighting, it becomes necessary to produce tubing that is not only uniformly and evenly coated but also has high stability and light output comparable to other types of lighting units. In order to produce coated tubes of high luminosity for cold cathode lighting, the method of using nitrocellulose and butyl acetate, or a similar lacquer, in combination with the powders should be employed. The tubes are then baked in a high temperature oven which eliminates the organic material, leaving a uniform coating of phosphor on the inside walls of the glass.

The cheaper and quicker process of coating, which omits the baking procedure, uses a quick-drying alcohol combined with the fluorescent powder. In this method, firm adherence of the powder to the glass is possible only if the powder is ball-milled to very fine particles before mixing with the alcohol binder. The excessive crushing of the fluorescent powder in this milling process is apt to disrupt the crystalline structure of the powder and hence reduce both its luminous efficiency and its stability. For this reason, the baked coating process, while more expensive, is considered superior to the alcohol evaporation process, especially for cold cathode tubing for illumination.

Hot Cathode Electrode Construction. A hot cathode electrode is shown in the cutaway diagram of a fluorescent lamp in Figure 15.4. The coil in the center consists of a double-coiled tungsten wire coated with a substance intended to reduce the electrode voltage drop. A second or two before the hot cathode lamp goes into operation current flows through the coil, causing it to glow at red heat. This heat is required to boil electrons out of the coated tungsten coil. Since the surface area of the coil is small, a high temperature is required to supply sufficient electrons for the discharge column and to reduce the voltage drop at the cathode. Hot cathode lamps flicker as they go on because of the delay involved in heating the coil. Once the lamp gets under way the normal tube current is sufficient to keep the coil at operating temperature. The starting operation puts a severe strain on the

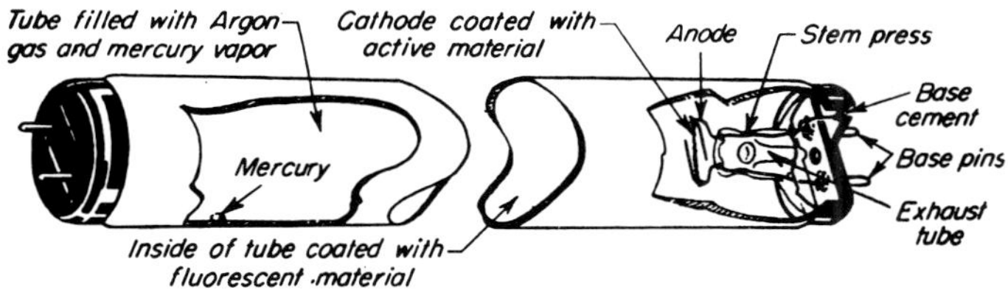

Figure 15.4. Cutaway diagram of standard preheat lamp. (General Electric Company.)

coated tungsten coil, and after a number of starts, the coil disintegrates, thereby terminating the life of the lamp.

While the electrode has a negative polarity, the hot coil emits electrons. At the same time, positive ions rush toward the electrode with tremendous momentum. If the coil were bombarded by all these high velocity ions, the disintegration would become very rapid and the lamp would fail. For this reason two heavy wires or anodes are placed on either side of the coil to act as targets for most of the ions, thereby reducing the number of ions that collide with the coil itself. When the electrode has positive polarity, it is bombarded by high velocity electrons. The same two heavy wires also protect the coil from excessive electron bombardment by serving as targets for the electrons.

To overcome the delayed starting characteristic of the preheated hot cathode lamp, the Slimline lamp and its electrode were developed. The coil of this electrode consists of heavier wire which stands more abuse during starting. As a result, the Slimline lamp is started by a high voltage kick, eliminating the need for preheating the tungsten coil and the consequent time delay in starting. When the circuit of a Slimline lamp is closed and the current begins to flow through the tube, the coil immediately develops a red-hot spot from which electrons are emitted. Figure 15.5 shows the difference between the pre-heat and Slimline electrodes. Although the Slimline electrode is sturdier, it also suffers during the starting operation, and burnouts may occur if the number of starts and stops is excessive.

The Slimline and preheated electrodes are both called "hot cathodes" since during operation they develop a hot spot having a temperature of approximately 950°C. Virtually all electrons are emitted from the hot spot. Currents as high as 1.5 amperes may be obtained from the preheated electrode and as high as 300 milliamperes from the Slimline electrode. The voltage drop of both cathode and anode varies between 14 and 20 volts. A limitation inherent in the design of these types of electrodes is the severe sputtering and disintegration of the coil during the starting operation which may cause the electrode, and consequently the

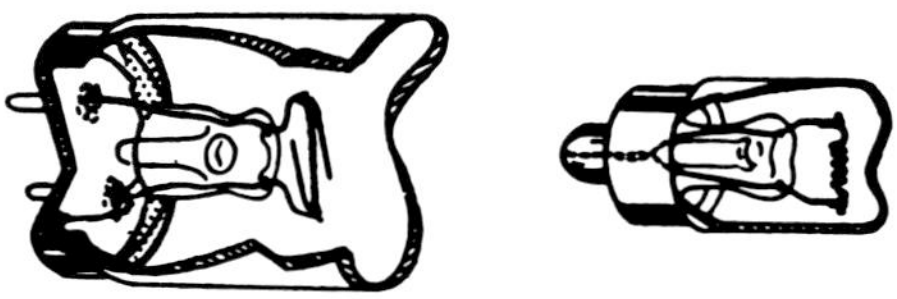

Figure 15.5. The Slimline cathode has a heavier filament than the preheat type on the left.

lamp, to fail after 1000 starts or less. For this reason, hot cathodes are unsuitable for any application requiring flashing or frequent starting. Under the most favorable practical conditions a burning life of 6000 hours may be expected from some types of hot cathode lamps, while under average conditions a useful life of 2000 to 4000 hours should be obtained.

Cold Cathode Electrodes. Cold cathode electrodes depend on a large emission surface area, rather than high temperature, for their operation. The emitting area in the hot spot of a hot cathode may be only a few thousandths of a square inch, while the emitting area of a cold cathode electrode carrying the same current may be approximately three square inches. This is the main difference between hot cathode lamps and cold cathode lamps—the construction of the electrodes.

A typical cold cathode electrode is shown in Figure 15.6. As a rule, these electrodes are similar in construction to those used for neon signs except that they are made with metal shells of larger size to carry higher currents. The metal shell is usually made of iron, with total impurities less than 0.1 per cent. The glass most commonly used for electrodes is lead glass, which can be sealed quite successfully to either lead or lime glass. Electrodes can be made using either nonex or pyrex for use with hard glass lamps. The inside of the shell is usually coated with an active emissive material which reduces the voltage drop at the electrode. The same coating material is used to coat the inside of a cold cathode shell as is used to coat the hot cathode tungsten coil.

While hot cathode electrodes operate on the principle of thermionic emission from the hot spot on the coil, cold cathode electrodes depend upon the electric field near the shell to emit electrons from the entire inner surface. The temperature of a shell reaches approximately 150°C under normal operation. The voltage drop at the electrode depends upon many factors, such as whether the shell is coated or uncoated, the size of the shell, the design of the shell, and the current through the tube. For an electrode operating at 120 milliamperes, and having a large surface area and coated with a good emissive coating on the inside of the shell, the voltage drop should vary between 90 and 100 volts. Smaller coated electrodes, operating on lower currents, may have a voltage drop that will vary up to 130 volts. Uncoated electrode shells will have an electrode drop approximately 80 per cent higher than that of the coated shells.

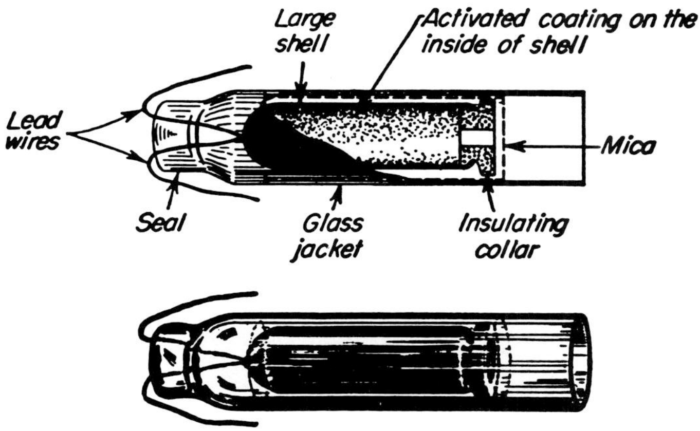

Figure 15.6. Typical cold cathode electrode.

Most of the cold cathode lamp installations that have been successfully made up to date use 100 to 120 milliamperes secondary current through the tube. Good installations, using up to 200 milliamperes, also have been made by companies using high quality production methods.

Although the hot cathode lamp has been universally accepted as the standard for the fluorescent lamp industry, it can be truthfully said that the cold cathode lamp offers many advantages. The cold cathode electrode is very rugged because of the use of a sturdy metal shell instead of a filament. Therefore it does not suffer during the starting process as does the hot cathode coil, and it can be flashed on and off without fear of damage. The life of a cold cathode tube is several times that of a hot cathode lamp because it is virtually impossible to burn out a properly processed cold cathode electrode. Generally a cold cathode tube must be replaced because of deterioration of the fluorescent powder rather than any defect in the electrode itself. In hot cathode lamps, nearly all replacements are necessitated by failure of the hot cathode coils. A life of 8000 to 10,000 hours may be expected from cold cathode lamps under average conditions, and 15,000 hours may be obtained under the best conditions. Even at this longer burning period, the only reason for replacing the tube may be that the light output of the fluorescent powder has fallen off to the point where the light output is below the requirement. But the old tube is still capable of burning. It is generally considered that the point at which it is economical to replace a lamp is when its light output drops to 65 or 70 per cent of the 100-hour value. In many cases where the quantity of light value is not very important, such as installations in bars, grillrooms, hallways, and auditoriums, the light output can drop far below the 65 per cent value without replacement.

Tubulated Electrodes. Several years ago manufacturers of the neon sign or press type of seal electrodes perfected a method for placing a tubulation on the end of the seal, as shown in Figure 15.7. This eliminates the need for sealing a tubulation on the

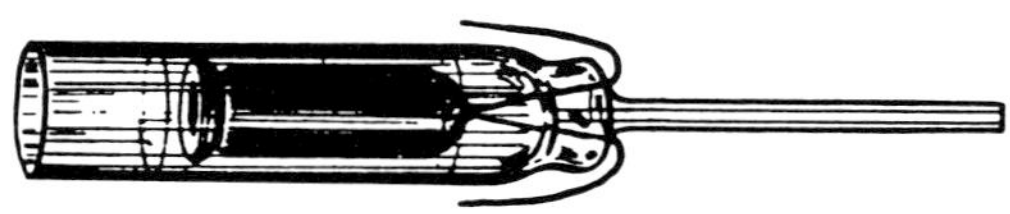

Figure 15.7. Tubulated cold cathode electrode.

tubing, which is both time consuming to fabricate and unsightly. The tubulated electrode is especially useful for cold cathode lamps used for lighting, since it helps produce a much more attractive finished lamp.

Ring-Seal Electrodes. Another means of improving the appearance of tubing is the use of ring-seal electrodes. The ring-seal stem has been used for a good many years in the manufacture of lamps and radio tubes. It is the type of seal that is now used in the manufacture of hot cathode lamps. Ring-seal stems come in pairs, one unit without a tubulation and the other unit with a tubulation sealed into electrode, as shown in Figure 15.8. In the modern method of making stems for lamps, tubulation glass is used on both ends for ease in handling. However, one stem has the hole blown through for pumping, and the other stem has a tubulation without the hole blown through the stem. This can be cut off as soon as it has served the purpose of ease in handling.

Recently some manufacturers of cold cathode lamps have adapted ring-seal stems for their fluorescent tubes. Figure 15.8 (*c*) shows a stem and cold cathode electrode in a cutaway section of a fluorescent lamp. One advantage of this type of a seal is that the fluorescent powder completely hides the electrode, giving the tube a very neat appearance.

Fluorescent Lamps. The three types of fluorescent lamps, cold cathode, preheat hot cathode, and Slimline hot cathode, are basically the same. They operate on the principle that a mercury discharge, emitting ultraviolet radiation, causes phosphors

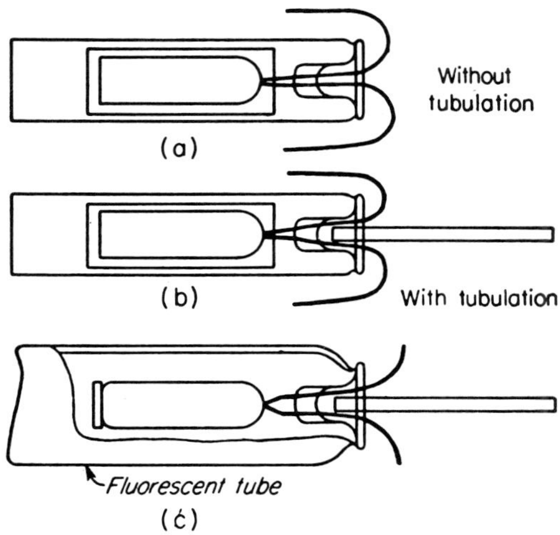

Figure 15.8. Three types of ring seals. (a) Electrode without tubulation. (b) Electrode with tubulation. (c) Tubulated stem and shell sealed directly to the fluorescent tube.

to produce visible light. Until recently, the same phosphors and mercury vapor pressure were used in all three types of lamps. The principal differences between the three lie in the electrodes, the length and diameter of the glass tube, the voltage required, and the intensity of current employed.

Preheat Hot Cathode Lamp. This lamp, developed around 1937, was the first popular fluorescent lamp. Despite the disadvantages of short life and delayed starting action, its superiority over the incandescent lamp and other commercial forms of lighting at that time was immediately recognized. With the constant improvements that have been made, it is widely used today. It is available in many sizes, the most popular of which is the 40-watt size.

The 40-watt (T-12) preheat hot cathode lamp is 48 inches long and 1.5 inches in diameter. The rated lamp current is 420 milliamperes, at which current the 4500 white lamp produces 525 lumens per foot when new. The operating voltage is 106 volts. The characteristics of preheat hot cathode lamps are shown in Table 15.1.

Slimline Hot Cathode Lamp. The Slimline lamp is a recent development. It was developed to overcome the disadvantage of a preheat starter and to increase the efficiency of the lamp by increasing its length. Slimline lamps are 42 to 96 inches long, the most popular being the longer lamp which can be contrasted

Table 15-1. Some Characteristics of Fluorescent Lamps*

Lamp watts†	Color of lamp	Lumens	Lumens per watt	Brightness, foot-lumens	Bulb diameter, inches	Length, inches‡	Lamp volts	Lamp current, amperes
8	3500 white	330	41	2,950	0.6	12	57	0.16
	4500 white	310	39	2,770				
15	3500 white	615	41	2,080	1.0	18	55	0.31
	4500 white	600	40	2,030				
20	3500 white	920	46	1,450	1.5	24	59	0.36
	4500 white	860	43	1,360				
30	3500 white	1470	49	2,260	1.0	36	98	0.35
	4500 white	1380	46	2,120				
40	3500 white	2300	58	1,750	1.5	48	106	0.42
	4500 white	2100	52	1,610				
	Daylight	1920	48	1,470				
	Soft White	1720	43	1,310				
100	3500 white	4200	42	1,850	2.1	60	71	.50
	4500 white	4000	40	1,750				

* General Electric Company.

† Add auxiliary watts for total.

‡ Length includes bulb and two standard lamp sockets.

with 100-watt preheat lamps having a maximum length of only 60 inches. The 96-inch (T-8) Slimline lamp with a diameter of one inch requires a starting voltage of 750 volts. A newly installed 4500 white lamp produces 260 lumens per foot, operating with a current of 120 milliamperes; and 382 lumens per foot, operating with a current of 200 milliamperes. Slimline lamps are being manufactured with diameters larger than one inch.

Cold Cathode Lamp. The cold cathode lamp contains electrodes having a large surface area which makes it a rugged and durable fluorescent lamp. Its life is independent of the number of starts and is much longer than that of a hot cathode lamp. This characteristic is particularly useful where frequent lamp replacement is difficult. An important advantage is that cold cathode tubes made by the average shop can be bent to any required shape, size, or length, whereas the hot cathode lamps are mass produced in standard sizes and shapes. The most popular cold cathode lamp is the 93-inch tube, one inch in diameter, used with an operating current of 100 or 120 milliamperes. Installations have been made using this lamp up to 200 milliamperes, and many manufacturers of cold cathode installations are working with currents of higher value. A 4500 white cold cathode fluorescent lamp operating on 100 milliamperes produces approximately 250 lumens per foot when new. Table 16.5 in Chapter 16 shows the light output, using various colors, diameters, and operating currents.

Circuits and Installations. The characteristics of fluorescent lamps necessitate electrical circuits inherently more complex than those employed with incandescent lamps. In order to realize fully the advantage of fluorescent lighting, the best design and materials must be used in the auxiliary equipment. Faulty ballasts, housings or sockets, wiring, or installation can easily nullify the high quality obtainable in fluorescent lamps.

Preheat Hot Cathode Circuit. Every preheat hot cathode circuit must perform three functions. (1) It must cause current to flow through the electrodes for a short time before voltage is applied to the lamp itself. This is necessary in order to bring the cathodes up to operating temperature. (2) It must produce sufficient voltage to start the discharge through the lamp. This may be accomplished by transformer action if the breakdown voltage required is higher than the line voltage. (3) It must provide impedance in series with the tubing to limit the lamp current to the operating or safe value.

The circuit shown in Figure 15.9 meets these requirements. The starters bridge the lamps for a short time after the switch is closed so that the ballast current flows through the tungsten-coil electrodes but not through the tube itself. When the coils are hot, the starters open, causing an inductive kick in the ballast to ignite the lamps. All the current must now flow through the tubes. The line voltage is boosted by the autotransformer which is a part of the ballast itself and shown in the left hand part of the unit. The two windings shown in the center of the ballast provide the necessary impedance to limit the current. In addition to performing these minimum functions, the ballast shown contains circuits which adjust the phase of the two lamps to eliminate stroboscopic effect. The power factor is also maintained at better than 90 per cent.

Slimline Hot Cathode Circuit. A Slimline circuit (Figure

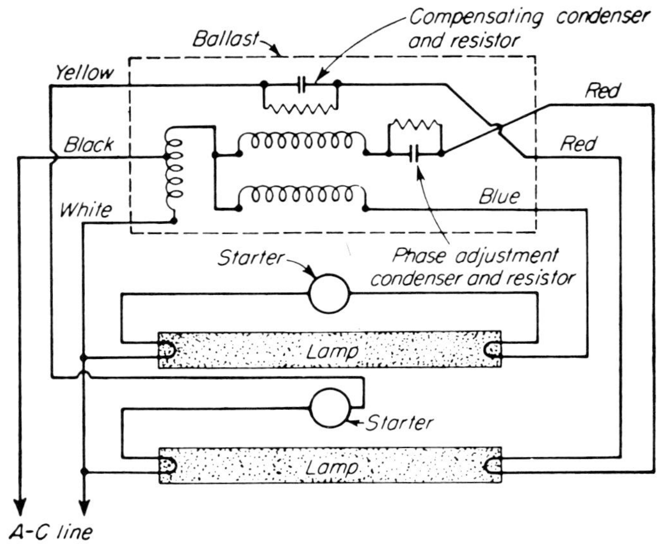

Figure 15.9. Typical two-lamp hot cathode circuit using a single
Figure 15.10. Slimline lamp circuits.

15.10) is simpler than a preheat circuit because no preheat starter is needed. The only requirements are that sufficient starting voltage be produced by transformer action and that enough impedance be placed in series with the lamps to restrict the lamp current to a safe value. Slimline lamps are usually operated from ballasts, having open circuit secondary voltages between 700 and 1000 volts. These ballasts can carry two lamps, each 96 inches long, and are corrected for power factor better than 90 per cent and for stroboscopic effect. The sockets used for Slimline lamps are designed so that the primary circuit is broken if a lamp is removed from the sockets. Because no preheat is necessary, Slimline lamps also can be used with high voltage transformers for series operation.

Cold Cathode Circuit. The requirements of a cold cathode circuit are the same as those of a Slimline circuit. Sufficient starting voltage must be available, and the operating current through the lamp must be limited by impedance in series with the tube. Cold cathode circuits have been developed into two distinct types, low voltage and high voltage, and each has its own particular advantages and disadvantages. This division resulted partially from the desire to adhere to the National Electric Code and other regulations.

Low Voltage Circuits. Circuits having a voltage under a thousand volts are classified as low voltage circuits, while those whose voltage is higher than a thousand volts are called high voltage circuits. Low voltage circuits are generally operated from ballasts, although some special low voltage circuits may employ transformers.

A cold cathode ballast is similar in many respects to the hot cathode ballast (Figure 15.11). An autotransformer boosts the

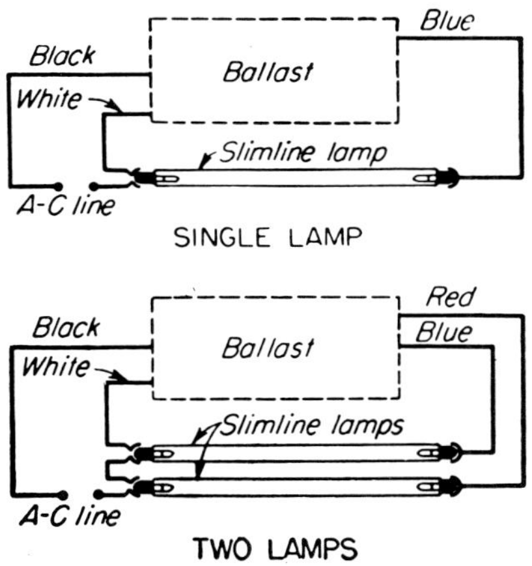

Figure 5.10. Slimline lamp circuits.

voltage to 950 volts in order to break down the gas column within the tube. After a 93-inch tube lights, the voltage falls to approximately 450 volts. The ballast has the correct impedance to limit the current from 100 to 120 milliamperes. A two-tube ballast circuit, shown in Figure 15.12, adjusts the phasing of the tubes to eliminate stroboscopic effect. In addition, the power factor is maintained at better than 90 per cent.

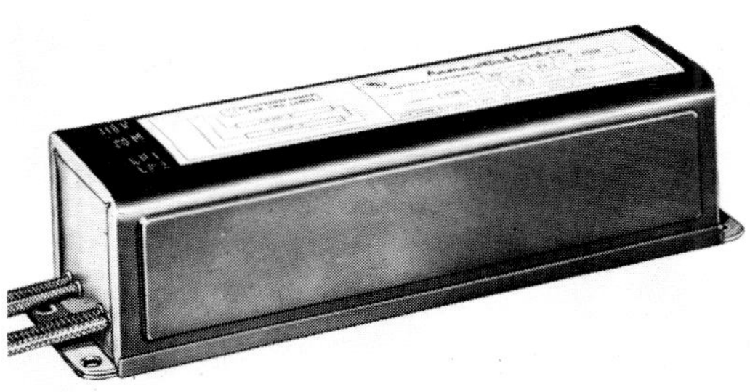

Figure 15.11. A typical cold cathode lamp ballast.

Numerous housings and sockets are available for low-voltage cold cathode circuits. A popular housing is shown in Figure 15.13. The wiring diagram for this type of housing appears in Figure

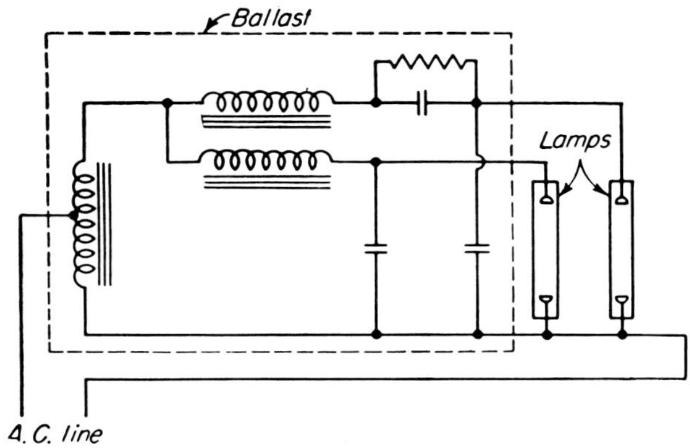

Figure 15.12. A cold cathode circuit using a single ***ballast for two*** *lamps.*

15.14. An important feature is that when the tube is removed from the housing a spring is released, which opens the primary circuit of the ballast. This prevents the possibility of anyone getting a shock by placing his finger in the empty socket. It is important that ballasts be mounted or fixed in position in an installation in such manner that sufficient ventilation is provided by air circulation. If ballasts or transformers get too hot, the potting compound within the unit melts or may expand, causing the ballasts to become defective.

High Voltage Circuits. A high voltage circuit contains several cold cathode tubes in series, powered by a high voltage transformer, as shown in Figure 15.14. If standard 93-inch tubes, 25 millimeters (one inch) in diameter, are used, the transformer will be properly loaded when it carries one tube for each thousand volts of secondary voltage. For these types of tubes, an operating current of 100 milliamperes, is generally employed, although lower currents are used with excellent results if a lower light value is desired. Many special installations have been made using higher operating currents. But, in general, the majority of cold cathode jobs installed use currents up to 120 milliamperes. The cold cathode high voltage transformers are available in standard sizes from 5000 to 15,000 volts secondary and up to 120 milliamperes secondary current. In general practice it has been found that a 12,000-volt, 120-milliampere transformer is the maximum secondary voltage transformer that should be used for best results. These transformers can be obtained with power factor corrected better than 90 per cent. There are available 120-milliampere transformers which are not power factor corrected. These uncorrected power factor transformers can be used if the local regulations permit or if they can be power factor corrected by the separate addition of the necessary capacitors connected to the primary side of the transformer.

Regulations require that the mid-point of all high voltage transformers for cold cathode illumination be grounded.

When high voltage is employed, there is a great tendency for arcs to occur to either ground or other points at the electrodes and housings. The higher the secondary voltage of the transformer, the more careful one must be in making proper connections and in the use of the insulators employed. Only the best housings and approved cable with proper insulation should be installed in a high voltage cold cathode lamp circuit. Permanent connections must be absolutely tight. Removable connections, such as the contact between an electrode cap and a spring, should be firm so

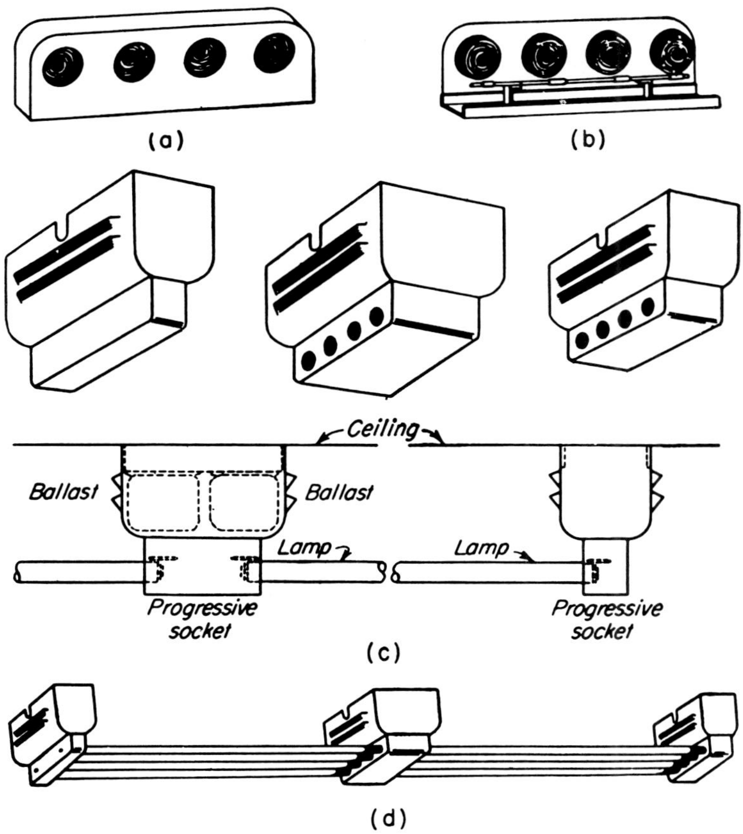

Figure 15.13. Cold cathode Progressive housings for low voltage (less than 1,000 volts). (a) Housing for four lamps with cover. (b) Same housing without cover. (c) Method of mounting housings on Royalux boxes. (d) Typical eight-tube installation.

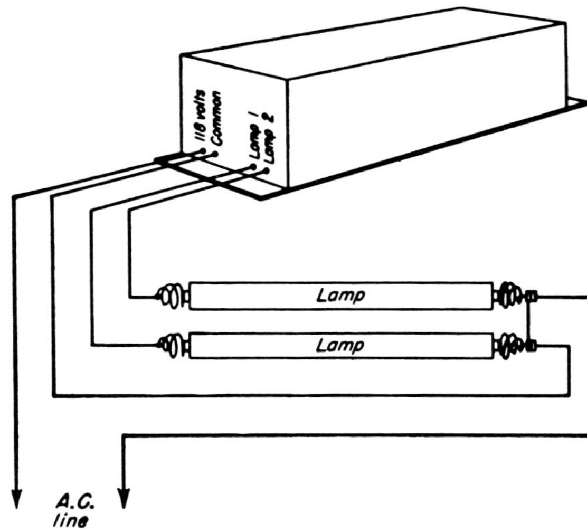

Figure 15.14. A two-lamp cold cathode schematic circuit using a ballast and Progressive-type housings. When a lamp is taken out of its socket, the primary circuit is automatically opened.

that there is no tendency for any arcing between two such contacts. Service organizations such as the Underwriters' Laboratories and the Canadian Standards Association test and approve the materials and finished units as to the kind of material used, the insulation qualities of a high voltage insulator, the suitability of certain cable, and any other tests of articles that may be required for high voltage circuits. But in the final analysis the greatest factor in the proper insulation of a high voltage circuit is the

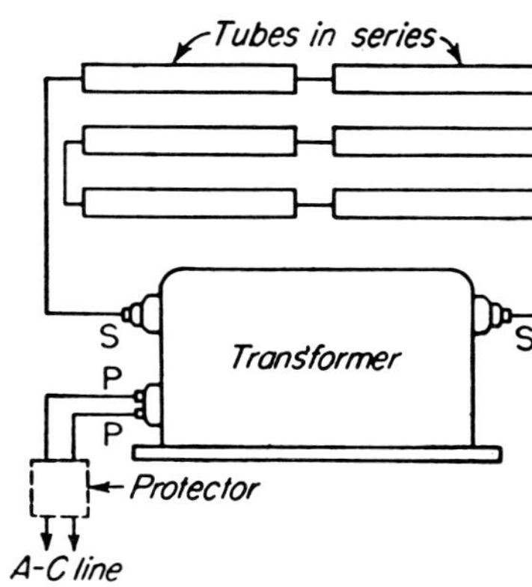

Figure 15. 15. A high voltage cold cathode lamp circuit.

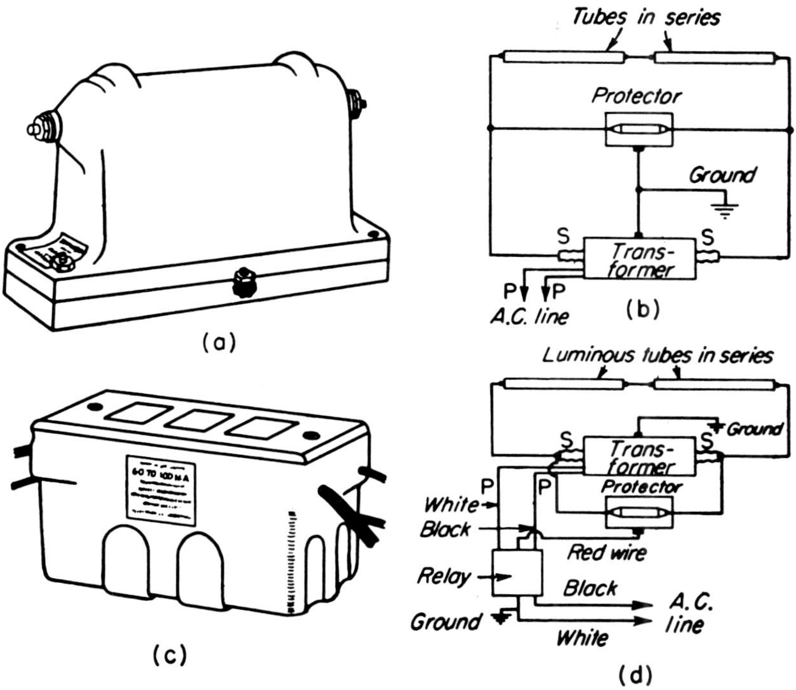

Figure 15.16. The protector (a) can be used by itself as in wiring diagram (b), or together with relay (c) as in wiring diagram (d).

method by which these materials are used and the method for connecting together all the units comprising a high voltage circuit, so that maximum insulation is provided against fire hazards and against human contact with live parts.

For example, a typical high voltage cold cathode circuit, using a 12,000-volt, 120-milliampere transformer, should be connected so that approved insulators are used, the high voltage cable suitably protected, and the transformer mounted in a metal case. This case should have a removable cover connected to the low voltage primary switch so that the primary circuit is broken if the cover is opened. The circuit should also have a suitable circuit breaker or protector to open the primary circuit if any trouble should occur in the secondary circuit, such as may be due to a defective tube.

There are a number of protectors on the market which can be used either for neon signs or for cold cathode lighting installations. Figure 15.16 (*a*) and (*b*) shows one type of protector which can be used by itself for neon signs; for cold cathode installations, it is used together with a relay, as in Figure 15.16

(*c*) and (*d*). When a break occurs in the secondary circuit, a bi-metallic or thermal switch is activated which in turn causes the relay to operate and open the primary circuit. When the trouble has been rectified, the protector will go back to normal position. Figure 15.17 shows another type of protector which

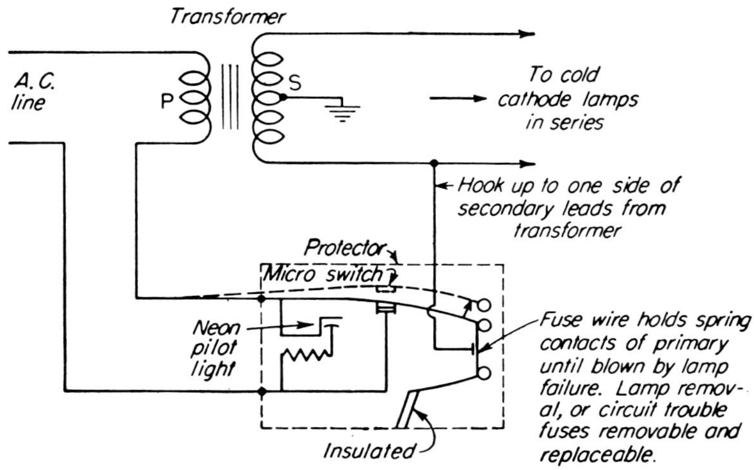

Figure 15. 17. Fuse-type protector.

operates on a fuse principle. Upon the opening or break of the secondary circuit, an arc occurs inside of the relay, across the fuse opening a microswitch in the primary circuit. The fuse is removable and replaceable, and as soon as the trouble has been rectified and a new fuse put in place, the circuit can be turned on once again.

A number of groups of housings are used for high voltage circuits for cold cathode. Figure 15.18 (*a*) and (*b*) shows two types of housings that have wide popularity. One type is used to connect two tubes in series. A tube is inserted on each side of this housing, and the electrical connection is made internally through a heavy phosphor-bronze spring. The spring, besides being the electrical connector between two tubes, also allows the tube to be pushed in so that it will fit easily into a similar housing at the other end. This places the tube in correct operating position so that both electrodes make firm contact with the respective springs. The internal construction of this housing is shown in Figure 15.18(*c*). For connection to the transformer, a housing of slightly different design is used shown in Figure 15.18(*d*). One end of the housing has a Fahnestock clip for electrical connection to the high voltage cable. After the cable connection is made, the Fahnestock clip and cable are pushed into the porcelain bushing.

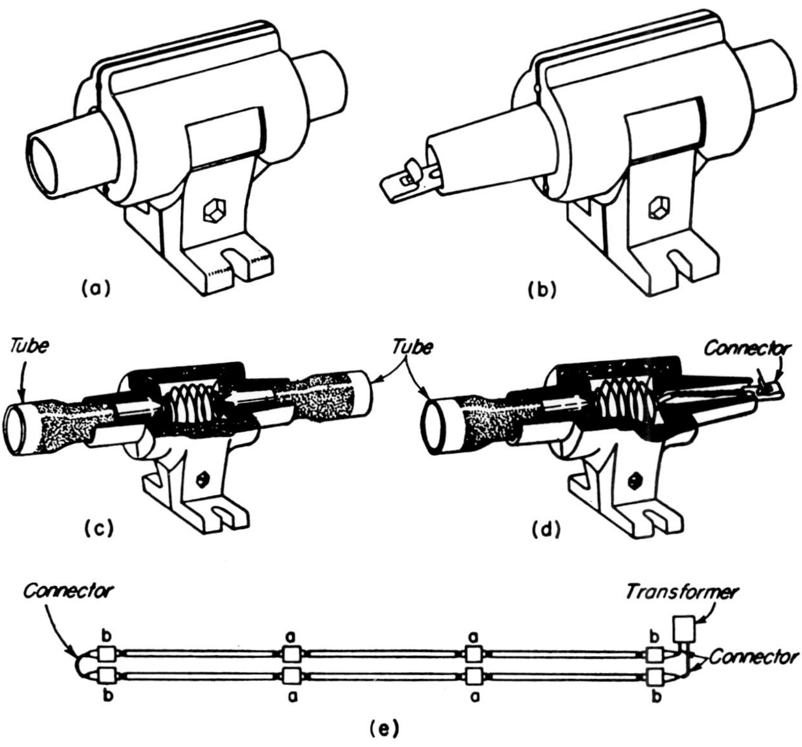

Figure 15.18. In-line housings for cold cathode lamps. Diagram (e) shows the use of housings (a) and (b).

Another group of housings for high voltage is shown in Figure 15.19. These housings are made in two parts. The tube is connected to a porcelain insulator that has a metal jack. The jack is snapped into place in the other porcelain socket which has already been fixed to either the ceiling or other permanent position.

One disadvantage of the two groups of insulators described above is that there will be a certain amount of dark space between lamps. This is true of any fluorescent lighting installations where straight lamps are used because the connecting sockets take up a certain amount of space when tubes are used in line. When an uninterrupted line of light is desired, the housings shown in Figure 15.20 (*a*) and (*b*) can be used. However, these housings have been designed for lamps that have the electrodes placed at right angles on each end of the tube, as shown in Figure 15.20 (*c*). The end housing employs a Fahnestock clip for connecting the high voltage cable to the transformer.

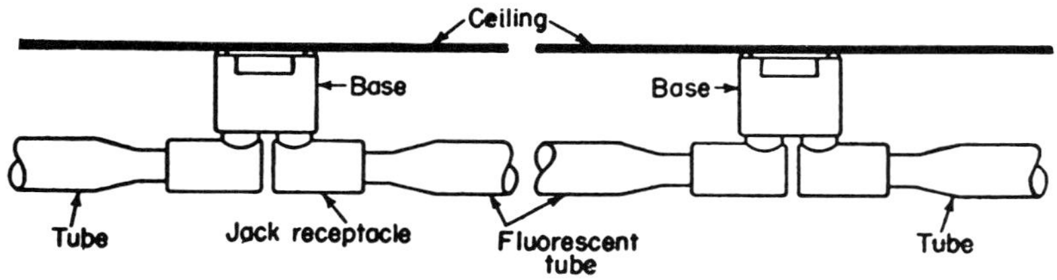

Figure 15.19. Jack and base receptacles for cold cathode lamps.

Caps and Terminals. In order to protect the electrode from breakage when it is placed into a housing, it is necessary to use a cap or terminal, as shown in Figure 15.21. The cap or terminal is placed on the end of the electrode and firmly fixed to the glass by a basing cement. Wires from the electrode are soldered to the metal terminal. There are two types of standard all-metal

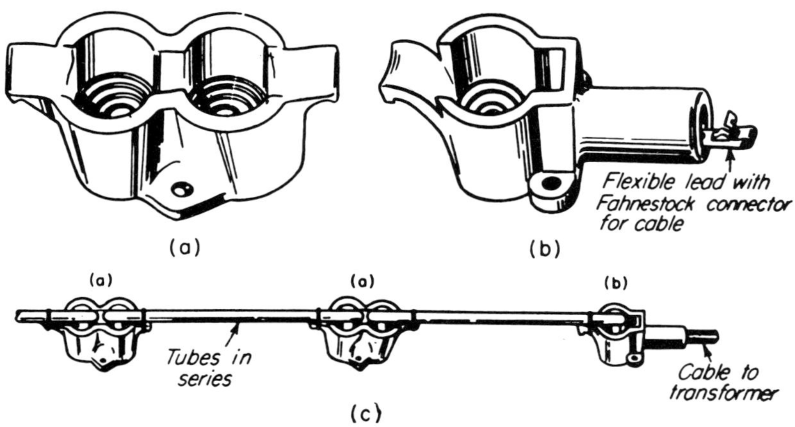

Figure 15.20. Housings for cold cathode lamps to produce an uninterrupted line of light.

caps used. One is a short unit to be placed on a standard electrode end. The other is a longer all-metal cap which is used for a tubulated electrode. The extra length of the cap protects that part of the tubulation which remained on the electrode end when the finished lamp had been sealed off from the pump.

Processing Lamps. In Chapters 8, 9, and 10 full details were given for pumping, bombarding, and filling luminous tubing for neon signs. In general, the same precautions and care must be taken for the processing of cold cathode fluorescent tubing. As a matter of fact, because of the higher currents used in cold cathode lamps, this tubing should be made under more exacting conditions. The processing consists of vaporizing all impurities in the

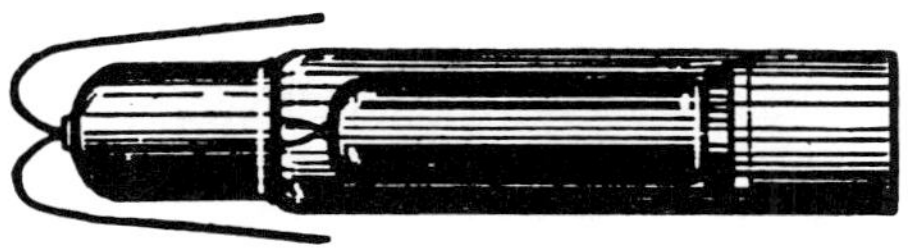

Figure 15. 21. Electrode with metal cap. Wires are removed after tube is processed.

tube, removing them in gaseous form, inserting the proper amount of rare gas and mercury, and sealing the tube off from the manifold. In order to vaporize impurities, it is necessary to heat all the glass to at least 200°C and all metal parts to 1000°C in a high vacuum. There are two general methods of accomplishing this.

Mass Production of Hot Cathode Lamps. Hot cathode lamps, manufactured in mass production, are usually processed in an automatic rotary exhaust machine. The lamps are placed in an upright position in the machine with the tubulation going down into a clamping device. The filament connections are then made, and the exhaust lever is opened. The lamps rotate in this machine through an oven which heats them to the proper temperature. At the same time, the filaments or electrode coils are heated to approximately 1000°C. The heating of the glass and the electrodes releases all impurities which are automatically pumped out by a group of pumping systems consisting of mechanical and condensation pumps in series. After the impurities have been thoroughly eliminated, a few millimeters of argon gas and several milligrams of mercury are inserted into the lamp and then it is sealed off. The caps are attached to each end and the wires from each electrode soldered to the metal pins.

The Production of Cold Cathode Lamps. Production of cold cathode lamps may be accomplished in one of two ways. They may be made in the average neon shop, using the same system that is used to process neon signs. This method uses internal bombarding by a high voltage transformer for heating both the glass and the electrodes. Because of the large volume of gases or impurities that must be eliminated from a cold cathode tube such as the standard 93-inch lamp, the pumping system must evacuate the tube to a high degree of vacuum in a minimum amount of time. This is of prime importance. This differs from the method employed in hot cathode lamps because human observation and control take the place of automatic machinery. It is therefore important that adequate instruments and a careful routine be employed by the man in the shop to ensure good results. A good pumping plan is shown in Figure 8.1. A good mechanical vacuum pump should be used in series with a condensation pump. It should be possible to achieve a vacuum of 0.1 micron in three minutes with no load on the system. The vacuum can be measured by any one of the gauges described in a previous chapter; of these the most popular are the McLeod gauge and the Pirani gauge. The Pirani gauge is more suitable for this type of plant because, as indicated in Figure 8.18, it not only measures a vacuum but also serves as an indicator of any leaks that may occur in a system. The system should have a leakage of no more than 0.5 micron per minute.

The bombarding transformer should be at least 7.5 kilovolt-amperes in size. For best results, a 12.5 kilovolt-ampere bombarder is recommended. The open secondary voltage should be approximately 15,000 volts.

The second method employed in processing cold cathode lamps, especially in large production, is the use of ovens and external bombarding methods. There is no standard method used, and each system has been designed and developed to meet particular requirements. In some systems the lamps are placed in an oven and heated to the proper temperature for eliminating all impurities from the glass tubing. This temperature would be approximately 400 to 425°C. After the tubes have been baked, the external bombarder coil is placed over each electrode and the electrode is heated to the high temperature required. In another system the cold cathode tubes are inserted in an oven, and at the same time, every electrode is encircled by an external bombarding coil fixed in place on the system. When the tubes have been heated to the proper temperature, a very large external bombarder is turned on, which supplies the necessary current to the copper bombarding coils to heat all the electrodes at the same time.

No matter which system is employed, necessary provisions must be made for inserting the proper amount of gas and the necessary quantity of mercury.

Type of Rare Gas Used. Fluorescent lamps may be filled with No. 50 (B-10), argon, krypton, or a mixture of krypton and argon. The kind of gas that is to be inserted and the amount of pressure used depends entirely upon the type of lamp and the system with which it is to be employed. On high voltage installations with many cold cathode tubes in series, No. 50 gas is the most satisfactory. This gas has a small quantity of neon added to the argon and tends to give a slightly more brilliant light output. The increase of resistance due to the addition of the neon is not of great importance when used in this system. However, when cold cathode tubes are manufactured for use on ballasts or low voltage transformers, then the resistance of the tube becomes an important factor and the total resistance of the gas column must be kept to a minimum. In this case only pure argon is used because it has a lower voltage drop than the No. 50 mixture.

Recently, hot cathode lamp manufacturers have been using krypton gas for many of their lamps. Krypton has a lower resistance than any of the other rare gases, with the result that the efficiency of the lamp has been increased. Also there has been a great deal of work done by using a combination of krypton and argon. Krypton by itself has the disadvantage of requiring higher temperature for starting than is required with argon. Also the efficiency of a lamp with krypton may suffer considerably at the lower bulb wall temperatures. This has been overcome somewhat by the addition of argon, giving a mixture that has approximately 50 per cent each of both krypton and argon.

The use of krypton gas or krypton-argon mixture in cold cathode tubes has been limited. A great deal of work is being done to determine whether there is a krypton-argon mixture of proper proportions that can be applied under the various conditions to which a cold cathode system may be subjected. Argon and argon-neon mixtures have proved to be best so far for cold cathode and neon sign work.

The quantity of gas that is used in tubes depends on many different factors. For small diameter tubing operating on high voltage systems a higher pressure of gas is used. For larger diameter tubing, such as 20 to 25 millimeters, a lower pressure of gas is used. Actually the quantity to be used should be low enough to give the lowest possible voltage drop across the tube and, at the same time, prevent the disintegration of the electrodes or the too rapid absorption of the gas because of too low a gas pressure. For example, in 25-millimeter tubing 93 inches long, the optimum pressure range is four to six millimeters of rare gas, taking all factors into consideration.

For hot cathode lamps where larger diameters are used, the quantity of argon gas inserted in a tube may be slightly less than four millimeters. It is understood that pressures as low as three millimeters of gas are used when either a krypton or a krypton-argon mixture is inserted.

Mercury. When mercury is used for fluorescent lamps in which the light output of a tube is of major importance, it is necessary to use only the highest quality and purity, that is available. As a matter of fact, only a minute quantity is inserted in a tube, and, therefore, the relative cost between mediocre and high quality mercury is slight.

The actual quantity to be inserted into a lamp is a difficult factor to determine. In the hot cathode lamps manufactured in mass production, the mercury is inserted sometime during the heating process so that the mercury becomes immediately vaporized throughout the tube. Also, as a rule, because of higher currents used, the ambient temperature of a hot cathode lamp is much higher than that of the same diameter cold cathode lamp. Therefore, the principal factor governing the quantity of mercury to be used is that there should be complete vaporization throughout the entire lamp length. This should be done using the smallest amount of mercury possible. The more mercury used, the greater the possibility of causing decrease of illumination after a length of burning time. The mercury tends to combine with any free oxygen that may be released, forming mercuric oxide which collects as rings near the electrodes and spreads inwardly toward the center of the tube. Also, excess mercury in the tube itself may form streaks throughout the lamp.

For approximate quantities of mercury to be used in tubing see page 164. The important factor is that the ambient temperature of the lamp should be such that any mercury within the tube is fully vaporized throughout its entire length. In many cases, a cold cathode tubing installation may be placed in sections where there are fans blowing on the tubes or other cooled areas not allowing the wall temperature of the lamp to reach the proper vaporization temperature. This may cause the tubes to burn dimly because of the fact that the mercury may not become fully vaporized throughout the tube. It is important that this factor should be taken into consideration when an installation is designed for cold cathode lighting. During operation, the bulb wall temperature should be between 90 and 130°F for best results.

Processing Routine. The following processing routine is recommended as a general procedure to obtain good results for manufacturing cold cathode tubes, using internal bombarding for heating up the glass and the electrodes.

1. Evacuate the tube until a discharge occurs when the bombarder switch is closed. When the discharge through the tube strikes, close the main pump stopcock immediately. If the discharge fades out, give the main pump stopcock one or two quick turns so that the discharge resumes.

2. Bombard until the glass temperature reaches 190°C. Some sort of precision device should be employed to indicate accurately the temperature of the glass while bombarding. A temperature indicator, as shown in Figure 9.4, is suitable. If no good means of measuring temperature are available, a piece of paper may be used as a very rough indication. Paper begins to get brown, on the average, at approximately 150°C and starts to turn black at 200°C. Of course, a lightweight paper will char faster than heavy paper. These figures apply to an average piece of newsprint.

3. When the glass reaches a temperature of 190°C, open the pump stopcock and continue bombarding until the electrodes are a bright red color. The entire surface of the electrode must get a bright red, because if any part of the electrode remains unheated, the electrode coating within the shell at that point will not be completely broken down. (Any electrode coating that does not break down completely on the bombarding table will later decompose at a slow rate while the tube is burning in the installation. The gases liberated by decomposition then collect in the tube and cause darkening.) By the time the electrodes are sufficiently heated, the glass temperature will have risen to between 200 and 225°C.

4. When the electrodes are thoroughly heated, stop bombarding and allow the tube to be evacuated.

5. Before the temperature drops to 150°C, the pressure should be one micron or less. If this condition cannot be obtained, the system pumps too slowly.

6. When the tube has cooled to approximately 80°C and the pressure reaches 0.5 micron, the tube is ready to be filled with the necessary pressure of rare gas and the proper amount of mercury.

CHAPTER XVI

Illumination Design For Fluorescent Installations

The term "illumination design" means the planning of a lighting installation so that it will provide adequate illumination for the intended use. Logically, such planning should precede the manufacture of the tubing in a cold cathode fluorescent system, and it should precede the installation of the equipment whether high voltage or low voltage tubing is used. Successful illumination design demands knowledge of the total light output from a given footage of tubing plus knowledge of the fixtures, room area, and absorption in attenuating or conserving the light. This chapter presents the basic procedure on which fluorescent illumination design is based.

Illumination Principles. The basic unit of light quantity is the lumen. This term has already been used in connection with luminous efficiency, which is measured in lumens per watt. In fluorescent tubing, the light output is given in lumens per foot. The total quantity of lumens produced by a given length of tubing is computed by multiplying the lumens per foot by the number of feet.

The intensity of illumination at a given surface, which is a measure of the utility of the lighting system, is measured in foot-candles and by a light meter (see Figure 16.1). A foot-candle is defined as the illumination resulting from one lumen falling uniformly over one square foot of surface. Thus, if we know the total number of lumens provided by the light source, and the total area on which it falls, we can compute the intensity of the illumination simply by dividing the lumens by the square feet. However, this process is possible only if the light falls uniformly over the surface. In nearly all practical cases this condition is not met. Rather, some of the light is intercepted by the fixture and is lost unless reflected back onto the surface to be illuminated. Moreover, most fixtures do not distribute the light evenly. Finally, the walls of a room reflect different amounts of light, according to their color and light absorption properties. Thus it is evident that calculation of illumination, based on the light source output in lumens and the area illuminated, must be approximate unless very elaborate measurements and computations are undertaken. Fortunately such approximate calculations suffice for general planning on a lighting system, especially since

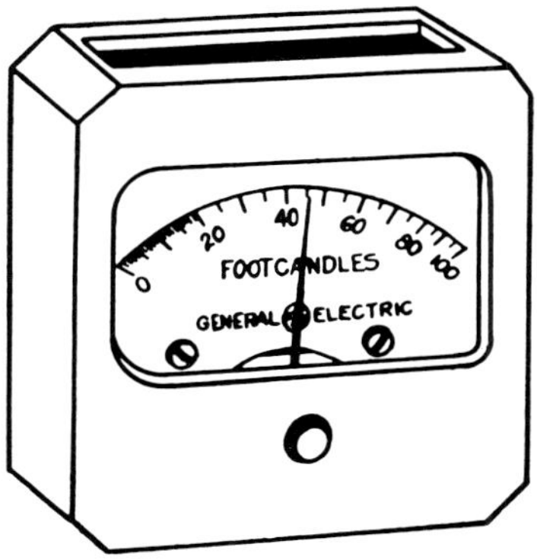

Figure 16.1. Meter for measuring light in foot-candles.

supplementary fixtures may be used to correct any deficiencies which may appear after the system is installed.

When the general level of illumination has been decided upon, and the number of feet of tubing or lamps required to produce the required level has been computed, the lighting design is only half complete. Thereafter it is necessary to lay out the position of the fixtures so that the light is properly directed to the "work area," that is, the surface in which the illumination is actually used. Moreover, the fixtures must be placed so as to avoid glare and to minimize sharp shadows. Fortunately, the fluorescent type of light is spread over a much larger area than that of an equivalent incandescent bulb, so it does not appear as bright to the eye, and objectionable glare is thereby rendered less likely. Moreover, the extended nature of the fluorescent tube or lamp makes it possible to distribute the light over a large area of the ceiling or walls of a room, so that sharp shadows cannot be cast.

It cannot be too strongly emphasized that the efficiency of a lighting installation depends not only on the intrinsic efficiency of the lamps, but also on the care with which the light is conserved by proper reflecting fixtures and directed to the region where the light is of use. This fact is taken into account in illumination calculations in terms of the so-called "coefficient of utilization," which specifies the degree to which the available light is actually applied to the work area. The coefficient may vary from as low as 10 per cent (90 per cent of the light not usefully employed) in high-ceilinged rooms with dark walls and inadequate fixtures to as high as 60 per cent in a low-ceilinged room with bright walls and well-designed fixtures.

General Procedure in Lighting Design. The procedure in designing a lighting system involves nine steps, as follows:

1. Determine the amount of illumination required for the tasks to be performed in the lighted area.
2. Select the type of lamp fixtures to be used.
3. Determine the light-reflecting or light-absorbing properties of the walls and ceiling of the room.
4. Make an allowance for the gradual loss of illumination due to dust on the lamps and fixtures, loss of light output of lamps, etc.
5. Calculate the total amount of light (lumens) which must be produced by the lamps to provide the necessary illumination, from the selected fixtures, in view of the properties of the room and the expected loss of illumination.
6. Select the color, diameter, and operating current of the luminous tubing to be used. From these data the luminous output in lumens per foot can be found.
7. From the lumens per foot provided and the lumens required, compute the number of feet of tubing required.
8. Design the layout of the tubing and placement of fixtures so as to conserve light, while meeting the overall requirements of the installation.
9. Design the electrical layout and connections, including length of each tubing section, and number of multiple or series circuits, and the transformers, ballasts, and switching devices.

Each of the above steps is described in detail in the following paragraphs.

Step 1. Illumination Requirements. The amount of light required for given tasks is given in Table 16.1[1].

The range of illumination from 10 to 100 foot-candles is usually obtained with general lighting fixtures, whereas supplementary lighting, closer to the work area, is used to cover the range from 50 to 1,000 foot-candles. Supplementary lighting may consist of extra lamps or spotlights.

Step 2. Selection of Fixtures. There are five classifications of light fixtures: indirect, semi-indirect, general diffuse, semidirect, and direct. Figure 16.2 shows the relative percentage of light directed upward and downward for these five types of fixtures.

[1]Data in this chapter are compiled from Cold Cathode Light Design. *Bulletin EN*-167, Sylvania Electric Products, Inc.

Table 16-1. Recommended Illumination Levels for Seeing Tasks*
(May include supplementary lighting)

Required foot-candles	Industrial	Offices, schools, etc.	Stores
10 (rough work)....	Loading Packing Shipping Washrooms	Auditoriums Cafeterias Conference rooms Passageways Reception rooms	Stockrooms
20 (casual tasks)....	Foundry work Processing Pressing-shearing Woodworking	Casual desk work Classrooms Filing Libraries	Circulation areas
50 (constant tasks)...	Automatic machinery General assembly Fabrication Sewing Typesetting	Accounting Bookkeeping Critical desk work Stenographic	General merchandising
100 (critical tasks)...	Fine assembly Fine machining Inspection Proofreading	Drafting Business machines	Counter displays Showcases Show windows Wall cases
200 (specialized applications)	Color inspection Extra-fine assembly Fine inspection	Color work	Display work Show windows
500 (specialized)....	Color identification	Color work	Display work Show windows
1,000 (specialized)	....	Hospital operating	Display work Show windows

* For complete data, "Illuminating Engineering Society Handbook," 1951, is recommended.

Experience has shown that offices, banks, schools, and other public buildings are best lighted by indirect or semi-indirect fixtures. Industrial areas are usually lighted by the direct method, since the high dark ceilings of most factory buildings make the indirect method impractical. In merchandising, all five types of fixtures may be used, depending on the type of work or display to be illuminated.

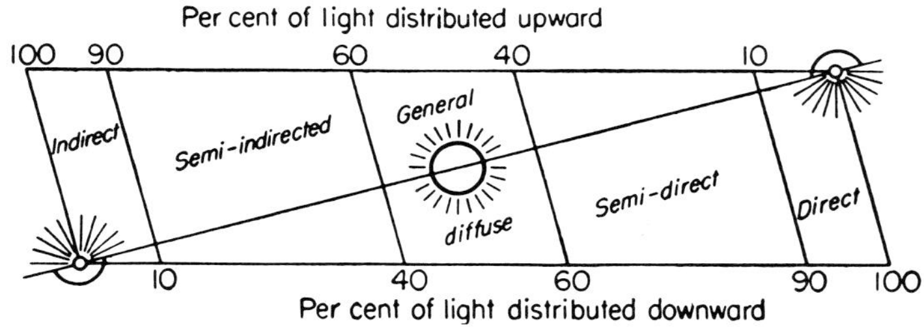

Figure 16.2. Light-distribution chart of lighting fixtures. (Sylvania Electric Products, Inc.)

If tubing is mounted on ceilings and connected to housings as described in the previous chapter, the same factors in lighting can be applied as with fixtures. In this case tubing can be classified as either direct or indirect. With direct lighting the tubing is mounted on the ceiling, which acts as a reflector. In indirect lighting the tubing may be mounted behind special troughs or coves.

In choosing and placing fixtures, care should be taken to meet the following general requirements: The fixture should harmonize as much as possible with its surroundings. Harsh shadows should be avoided. The distribution of light in the horizontal and vertical plane should be suitable. Intense localized areas of light, such as glare and reflected high lights, should be avoided. Finally, the fixtures chosen should be easily cleaned and maintained, since a poor fixture in this respect encourages poor maintenance with consequent excessive loss of light during the life of the installation.

The fixture types shown in Table 16.7 have widely different properties with respect to the distribution of light upward and downward. The light-distribution diagrams shown in the table should be studied in choosing the fixture type.

Step 3. Coefficient of Utilization. The effect of the room dimensions and the wall and ceiling coatings is expressed in the "coefficient of utilization." This is the ratio of the number of lumens reaching the working surface to the total number of lumens generated by the lamps. This coefficient must take into account the efficiency of the fixture, its light distribution curve, its height above the floor, the proportions of the room, and the reflection properties of the walls and ceiling. All these factors are shown in the Table of Coefficients of Illumination (Table 16.7). These tables are divided into sections according to the type of fixture, illustrated at the left. The light distribution curve of the fixture

Table 16-2. Reflection Factors of Various Materials and Surfaces*

Material	Reflection factor, per cent	Material	Reflection factor, per cent
White plaster	90–95	Polished gold	50–55
Polished silver	90–92	Alabaster veined	49–67
Alzak	85	Monel metal	49–55
White blotting paper	82	Limestone	35–58
White cartridge paper	80	Polished marble	30–70
Oxidized or etched aluminum	70–89	Tracing cloth	30
Mirrors, silvered	70–85	Impregnated marble	27–54
Highly polished brass	70–75	Colored alabaster	27–29
Ordinary foolscap	70	Light oak finish	25–35
Glazed white porcelain enamel	65–77	Gray cement	20–30
Mat white porcelain enamel	60–83	Granite	20–25
Chromium	63–66	Macadam road	12
Polished aluminum	62	Red brick	10–15
Terra cotta, white and cream, smooth and mat	60–80	Dark oak finish	10–15
Speculum metal	60–80	Mahogany finish	6–12
Highly polished copper	60–70	Walnut finish	5–10
Aluminum paint	60–65	Black paper	5
Highly polished steel	60	Black cloth	1.2
Caen stone	55–65	Dead-black paint	1
Chromium-nickel (stainless) steel	55	Black velvet	0.4

* *American Architect*, November, 1934 (American Institute of Architects, File No. 31-F).

is shown, and the relative proportion of light directed upward and downward is shown. Below the distribution diagram in each case are shown representative values of maintenance factors, for good, average, and bad maintenance (see the following Step 4).

To find the coefficient of utilization for a given fixture it is necessary to know the room index, which takes into account the geometric proportions of the room and location of the fixtures, and the reflection factors of the walls and ceilings, expressed in per cent. The latter figures must be estimated from the color and brightness or darkness of the wallpaper and paint. For simplicity in calculations of lighting system, typical reflection factors for ceilings of 50 and 75 per cent are used, and for walls 10, 30, and 50 per cent are used. Actual reflection factors of materials, surfaces, and painted surfaces are shown in Tables

16.2 and 16.3. As a general rule, use a reflection factor of 75 per cent for light-colored ceilings and 50 per cent for dark colors. For walls use 50 per cent for light colors, 30 per cent for medium colors, and 10 per cent for dark colors.

The room index, identified in the table by the letters A through J, is given in the table of room indexes, Table 16.8. In the latter table, the width and length of the room are first located at the left, and the room index is found under the ceiling height (semi-indirect and indirect lighting) or mounting height (direct and semidirect lighting).

In considering the room dimensions, it should be noted that the efficiency decreases as the index goes from A to J and that, in general, the larger the room, the more efficient the lighting design. This follows from the fact that in a large room the absorption of light by the walls and ceiling has relatively small effect.

Table 16-3. Reflection from Painted Surfaces*

Color	Reflection, per cent	Color	Reflection, per cent
Flat white	75–85	Spanish or bright orange	35–40
Gloss white	75–80	Medium brown	25–35
Cream	65–75	Dark gray	20–25
Bright yellow	60–70	Cardinal or dark red	20–25
Aluminum paint	60–65	Willow or medium green	15–25
Buff	52–56	Olive green	12–15
Very light blue	45–55	Dark brown	8–10
Apple or pale green	40–55	Spanish or bright blue	5–10
Light gray†	35–50		

* *American Architect* (American Institute of Architects, File No. 31-F).

† Grays vary remarkably, depending on the way they are prepared. A gray made by mixing lampblack with white paint has a low coefficient of reflection. A gray made by mixing red and green paint with a white base has a relatively high coefficient of reflection and is known as a warm gray.

Step 4. Maintenance Factor. It is not sufficient to design a lighting system on the basis of the light produced when the installation is new, since a number of factors conspire to reduce the useful illumination as time goes on. Primary among these is the accumulation of dirt and dust on the lamp and the fixture itself. Of equal importance, but often neglected in computations, is the effect of dirt and dust on the walls and ceiling of the room. Finally, every lamp, cold cathode, hot cathode, or incandescent, has a definite loss of output with life.

Maintenance is particularly important in cold cathode installations, since this type of lamp has long life and the total accumulation of grime during the life of the lamp is correspondingly great. The manufacturer should accordingly instruct the customer in the paramount importance of regular cleansing of lamps, fixtures, and walls. Figure 16.3 shows the progressive loss of

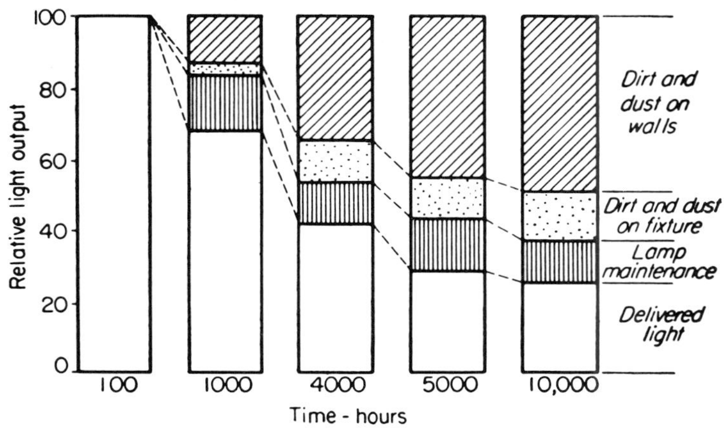

Figure 16.3. Chart showing that loss of light is due principally to dust and dirt. (Sylvania Electric Products, Inc.)

light output from 100 to 10,000 hours in typical installation. The final light output is only 26 per cent of the initial value, and two-thirds of the lost light is accounted for by dirt and dust on the lamps and on the walls and ceiling of the room.

The letters G, A, and B in the Table of Coefficients of Utilization (Table 16.7) show, respectively, the average loss of light due to the above causes when the maintenance schedule is G or good (frequent cleaning of luminaires and replacement of lamps as required), A or average (lamps only in case of failure), or B or bad (luminaires never cleaned).

Step 5. Computation of Lumens Required. From the number of foot-candles specified in Step 1, the area to be illuminated (usually floor area), the coefficient of utilization in Step 3, and the maintenance factor in Step 4, we compute the total number of required lumens by the following equation:

$$\text{Required lumens} = \frac{\text{foot-candles} \times \text{area in square feet}}{\text{coefficient of utilization} \times \text{maintenance factor}}$$

Step 6. Choice of Tubing. The available colors in cold cathode tubing have been discussed in earlier sections of this book. From the standpoint of lighting design, the color chosen should be coordinated with the amount of light required and the length of tubing which is permissible under the cost limit. Table 16.4 summarizes the color specifications with particular reference to interior illumination.

The diameter of the tubing depends on the type of fixture and the required illumination per foot. Large diameter tubing has lower lumen output per foot and higher efficiency than small diameter tubing. Where the tubing is exposed directly to the eye, large diameter low current tubing should be used to avoid glare. When the tubing is in a fixture which shields it from the eye, small diameter high current tubing may be used.

Table 16-4. Colors Available for Fluorescent Tubing

Color name	Resultant color when pumped with—		Suggested color uses
	Argon and Mercury	Neon	
Coated tubing using clear glass			
Daylight	Daylight*		Daylight—cold effect
		Coral	Decorative
4500 white	4500 white*		Sunlight—cool effect
3500 white	3500 white*		Good interior color—warm tone
Warm White	Warm White		Blends well with incandescent
Soft White	Soft White*		High red component—emphasizes warm colors
Pink-white	Pink-white		Applicable where definite pink color needed
Blue	Blue*		High brightness; pure blue color
		Rose-pink	Decorative
Green	Green*		Highest brilliancy color
		Amber-gold	Good neon gold color
Coated tubing using colored glass			
Noviol glass, green coating	Deep green		Approaches pure green color
Noviol glass, white coating	Yellow-gold		Bright canary yellow
Ruby glass, white coating	Soft red	Ruby rose	Signal red; high-brightness red; bright ruby red

* These colors will match standard fluorescent lamps when properly processed by introducing appropriate argon gas pressure and when operated at correct current density.

Table 16-5. Light Output of Fluorescent Tubing*

(Lumens per foot)

Diameter of tubing, millimeters	Operating current, milliamperes	Color of tubing			
		Daylight	3500	4500	Soft white
12	25	125	140	135	110
	50	230	250	240	200
15	25	100	130	120	90
	50	185	230	220	150
18	25	90	120	110	80
	50	160	215	195	140
	100	285	360	340	230
20	25	85	105	100	75
	50	150	185	170	130
	100	260	320	310	210
	120	305	370	350	260
25	25	75	85	80	60
	50	120	150	130	105
	100	235	265	250	190
	120	270	310	285	215

* Because of many variables in the manufacture of tubing, the table gives approximate values and can serve only as a guide in designing lighting installations. The variables may be (1) type of phosphors used, (2) variation in diameter and thickness of wall in glass tubing, (3) kind of rare gas and pressure used in finished tube.

All manufacturers of cold-cathode tubing should work out tables giving the lumens per foot output of their tubing. Accurate measurements and data can be furnished by service organizations such as the Electrical Testing Laboratories of New York City or similar testing laboratories.

Step 7. Computation of Length of Tubing. When the color, diameter, and current have been chosen, the light output in lumens per foot can be determined by reference to Table 16.5. This table shows the lumens per foot output for various operating currents and diameters of four widely used white tubing (Daylight, 4500 white, 3500 white, and Soft White).

To find the footage required, we devide the total lumens required (Step 5) by the lumens per foot of the selected tubing.

$$\text{Footage} = \frac{\text{total lumens required}}{\text{lumens per foot of tubing}}$$

Step 8. Layout of Tubing. The layout of the tubing involves the spacing of the fixture or single rows of tubing. Table 16.6 gives the suggested mounting heights, spacing, and suspension heights for various types of direct and indirect fixtures. It should be noted that these are the maximum spacings and that in any event the spacing should not exceed the mounting height. A fixture next to a wall should be mounted at a spacing equal to one-third of the distance between fixtures.

Table 16-6. Suggested Spacings for Luminaires, Feet*

Mounting heights above floor	Type of luminaire			
	Direct, semidirect, and general diffuse lighting		Indirect and semi-indirect	
	Spacing between luminaires		Spacing between luminaires	Suspension length of luminaires
	Diffused	Concentrating		
8	7	2½	9	1–3
9	8	3	10	2–3
10	9	3	10	2–3
11	10	4	12	2–4
12	11	4	14	3–4
13	12	5	14	3–4
14	12	5	15	3–4
15	12	6	18	3–4
16	12	6	20	3–4
18	18	8	20	4–6
20	20	8	20	4–6
22	20	10	22	4–6
24	23	10	24	4–6
26	25	10	26	4–6
28	30	10	28	4–6
30 and up	30	15	30	4–6

* Sylvania Electric Products, Inc.

The recommended spacing should be laid out on a plan showing the position of each fixture and the total number of fixtures required. The number of fixtures times the number of feet per fixture should be equal to the total footage computed in Step 7.

Step 9. Electrical Design. The electrical design starts with the choice of length of tubing sections. Cold cathode tubing is ordinarily made up in 93-inch lengths, although shorter lengths are used in many cases. In general, the longer the length, the higher the efficiency. However, too long sections of tubing should be avoided since they are difficult to install and maintain.

When the length of tubing section has been decided, the number of tubing sections is computed from the total footage required. It is then necessary to decide on the electrical connections to handle the given number of sections. Two methods are in general use, the multiple (parallel or shunt) connection and the series connection.

In the multiple-circuit, cold cathode lighting is similar to hot cathode lighting. Separate ballasts are needed for each lamp or two-lamp section. Series operation, shown in Figure 16.4, re-

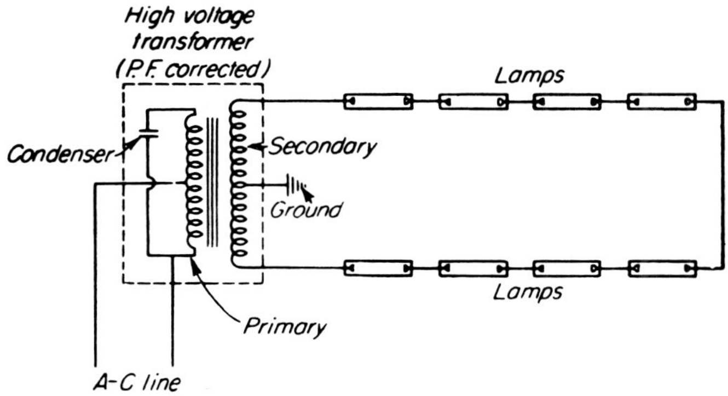

Figure 16.4. Cold cathode tubing in series operation.

quires a higher voltage transformer and is similar to the electrical layout of a luminous tube sign. When using the series connection, allowance for electrodes must be made on the basis that each pair of electrodes consumes the equivalent of one foot of tubing.

When the number of sections to be operated from one transformer is decided, reference to Table 6.2 in Chapter 6 will permit the choice of the required transformers.

Example of Calculation of a Cold Cathode Tubing Installation. Illuminate an office space requiring critical desk work, 50 by 100 feet with a 12-foot ceiling height, light walls (reflection factor = 50 per cent), light ceiling (reflection factor = 75 per cent), light desks, and average cleaning maintenance schedule. Make recommendation on the basis of series-type installation.

Step 1. Prescription of Lighting Requirements. From Table 16.1, this office space requires 50 foot-candles for very bright illumination.

Step 2. Select Type of Lighting System to Be Used. Since work is quite critical and eye comfort important, recommend a semi-indirect fixture such as that shown as fixture *(i)* in Table

16.7, page 288, having a distribution of 50 per cent upward and 30 per cent downward. A fixture of this type is built for use with ballasts. However, for the purpose of calculating this example, it is assumed that this fixture is designed for high voltage tubes in series operation.

Step 3a. Determine the Room Index. From Table 16.8 a room that is 50 feet wide × 100 feet long and has a ceiling or mounting height of fixture from 10 to 12 feet (semi-indirect or indirect) has a room index of A.

Step 3b. Determine the Coefficient of Utilization. From Table 16.7 with ceiling reflection factor of 75 per cent and wall reflection factor of 50 per cent, with a room index of A, the coefficient of utilization is determined to be 0.54 with the fixture *(i)* that has been selected.

Step 4. Estimate Maintenance Factor. From Table 16.7, under average conditions A the maintenance factor for fixture (i) is 0.65.

Step 5. Calculate the Total Lumens Required.

$$\text{Number lumens} = \frac{\text{foot-candles} \times \text{area}}{\text{utilization coefficient} \times \text{maintenance factor}}$$

$$= \frac{50 \times 50 \times 100}{0.54 \times 0.65}$$

$$= 712{,}000 \text{ lumens}$$

Step 6a. Select the Color of Tubing to Be Used. Since this is an office application where color is not an important factor as long as it is white, we will use tubing of greatest lumen output. Specify 3500 white tubing.

Step 6b. Select the Diameter and Operating Current of the Tubing to Be Used. Since this is a shielded application, we can use high brightness tubing. However, economics dictate that we also obtain high lumen output per unit of tubing with good life characteristics. So we will standardize on 93-inch nominal length sections of tubing which have a luminous length of seven feet, five inches, the remainder of the tubing being opaque owing to the presence of the electrode shell and the lamp holder. The chart below shows the tubing choices available from Table 16.5, when the 3500 white tubing is operating at 100 milliamperes. (Transformers are rated by their short-circuit current which is 20 per cent higher in value than the operating current. To give 100 milliamperes, the transformer should be rated at 120 milliamperes.)

Tubing Size, Millimeters	*Lumens per Foot at 100 Milliamperes*
18	360
20	320
25	265

We discard the 18-millimeter tubing because this size is small in diameter for standard 100-milliampere operation. The 25-millimeter tubing does not give high enough lumen output to justify its use in a shielded application. Therefore the choice is 20-millimeter tubes operated at 100 milliamperes.

Step 7. Determine the Number of Feet of Tubing Required.

$$\text{Feet tubing required} = \frac{\text{total lumens required}}{\text{lumens per foot of tubing}}$$

$$= \frac{712{,}000}{320} = 2{,}220 \text{ feet}$$

or in the 93-inch section (7 foot, 5 inch luminous length)

$$\text{Number tubes required} = \frac{2{,}220}{7 \text{ feet, } 5 \text{ inches}} = 300$$

Step 8. Lay Out the Tubing. Applying the data in Table 16.16, we can see that the tubing distribution should be as follows:

$$\text{Area per foot tubing} = \frac{\text{total area}}{\text{feet tubing required}}$$

$$= \frac{50 \times 100}{2{,}220} = 2.25 \text{ square feet}$$

or

$$\text{Area per unit length (8 feet)} = \frac{\text{total area}}{\text{number lamps required}}$$

$$= \frac{5{,}000}{300} = 16.7 \text{ square feet}$$

If we construct four lamp fixtures, the area per fixture will become: $16.7 \times 4 = 66.4$ square feet, or 75 fixtures will be required for 300 tubes.

In an area 50×100 feet, we can place 12 of the eight-foot fixtures so as to form a continuous row 96 feet long. In order to obtain approximately 75 fixtures, six rows will be required. Therefore the recommendation becomes 72 fixtures run in continuous rows of 12 fixtures each and these continuous rows spaced eight feet, four inches apart. The other three fixtures can be mounted in the most convenient space available. The fixtures should be suspended at least two feet, six inches below the ceiling, and since we have a 12-foot ceiling height, the brightness pattern on the ceiling will be improved if we suspend them three feet below the ceiling.

Step 9. Electrical Specification. The following electrical specifications are also required:

1. The number and type of transformers that will be required for the job.
2. The amount of power to be consumed in order to calculate the size of the wiring, distribution of the power, and the necessary switches, fuses, and other electrical units.

Transformers. Having determined the number of lamps and layout of these lamps in the room, the number of transformers required can be determined readily. NESA (National Electric Sign Association) has recommended that a transformer for cold cathode lighting should be limited up to and including 12,000 volts. Therefore we will use the largest permissible size, which is a 12,000-volt transformer. From Table 6.2, in Chapter 6, we determine that a 12,000-volt transformer operating at 100 milliamperes will carry 70 feet of 20-millimeter mercury tubing.

Summing up all the available information that we have so far, we obtain the following:

1. Diameter of lamp=20 millimeters
2. Length of each lamp=93 inches
3. Number of lamps=300
4. Total length of tubing=approximately 2,220 feet
5. Type of transformer to be used=12,000 volt, 120 milliamperes (short-circuited current)
6. Number of feet per transformer=70 feet
7. Number of eight-foot lamps per transformer=8 (Allowance of six feet is made for the electrode drop of the lamps.)
8. Number of transformers required=38 (Actually only 37 of the 12,000-volt transformers and one 6,000-volt transformer are required. The 6,000-volt transformer will light 32 feet, or four of the eight-foot lamps.)

Power Required. It is the custom to use only power factor transformers, corrected to better than 90 per cent, for interior lighting installations and especially when the power consumption is high. A 12,000-volt, 120-milliampere power factor transformer will require approximately 595 volt-amperes. It will require 5.2 amperes and will consume 550 watts. A 6,000-volt, 120-milliampere power-factor transformer will require approximately 310 volt-amperes. It will require 2.2 amperes and will consume 240 watts. Therefore, 37 of the 12,000-volt transformers and one 6,000-volt transformer will require:

Approximately 22,325 volt-amperes

Approximately 195 amperes

Approximately 20,590 watts

Conclusion. Any space can be properly illuminated by following, step by step, all the factors illustrated in the example.

Most cold cathode installations will use fixtures of type *(d)*, illustrated in Table 16.7. These fixtures may consist of sockets only which can be mounted directly onto the ceiling, or they may be complete units which are designed to be suspended by suitable supports at any height desired from the ceiling. With the cold cathode lamps completely exposed, practically all the light is concentrated downward toward the worktables.

To those who are interested in all phases of lighting installations, the author recommends "Illuminating Engineering Society Handbook," which is comprehensive and complete.

Table 16-7. Coefficients of Utilization*

Ceiling reflection factor, per cent		75			50		
Wall reflection factor, per cent		50	30	10	50	30	10
Cold-cathode fixture types†	Room index	Coefficients of utilization					
(a) 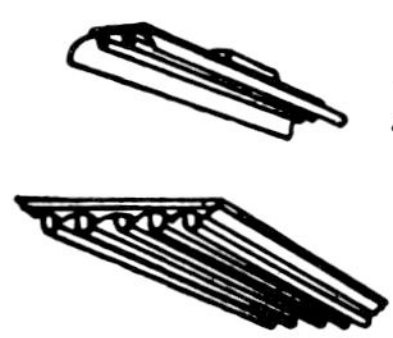 G=.70 A=.60 B=.55; 0/80	J	0.36	0.29	0.24	0.35	0.29	0.24
	I	0.45	0.39	0.35	0.44	0.38	0.34
	H	0.49	0.44	0.40	0.48	0.44	0.40
	G	0.53	0.48	0.44	0.52	0.48	0.44
	F	0.56	0.51	0.47	0.54	0.50	0.47
	E	0.61	0.57	0.53	0.60	0.56	0.53
	D	0.66	0.62	0.58	0.64	0.62	0.58
	C	0.68	0.65	0.62	0.66	0.63	0.62
	B	0.72	0.69	0.66	0.70	0.67	0.66
	A	0.74	0.70	0.68	0.72	0.69	0.67
(b) 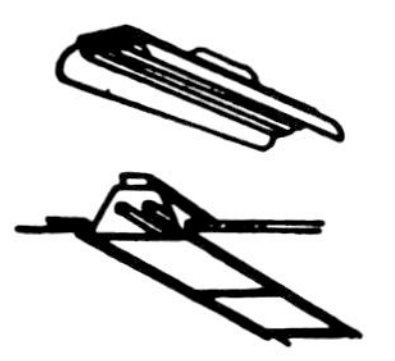 G=.70 A=.60 B=.55; 0/70	J	0.32	0.28	0.25	0.32	0.28	0.25
	I	0.40	0.36	0.34	0.39	0.35	0.33
	H	0.43	0.39	0.37	0.42	0.39	0.37
	G	0.46	0.43	0.41	0.45	0.43	0.41
	F	0.48	0.45	0.43	0.47	0.45	0.43
	E	0.52	0.50	0.48	0.51	0.49	0.47
	D	0.56	0.54	0.52	0.55	0.53	0.51
	C	0.57	0.55	0.53	0.56	0.54	0.52
	B	0.60	0.58	0.56	0.59	0.57	0.55
	A	0.61	0.59	0.57	0.60	0.58	0.57
(c) 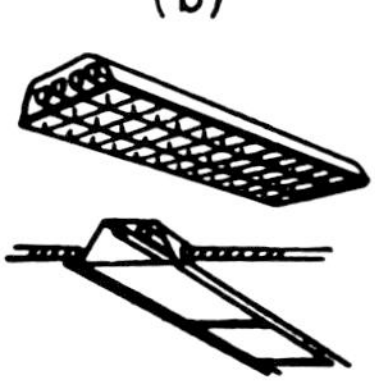 G=.70 A=.60 B=.50; 0/60	J	0.30	0.26	0.24	0.29	0.26	0.24
	I	0.37	0.34	0.32	0.36	0.33	0.31
	H	0.40	0.38	0.36	0.39	0.37	0.35
	G	0.43	0.40	0.39	0.42	0.40	0.38
	F	0.45	0.43	0.40	0.44	0.42	0.40
	E	0.48	0.46	0.44	0.47	0.46	0.44
	D	0.52	0.49	0.47	0.50	0.49	0.47
	C	0.53	0.51	0.49	0.52	0.50	0.49
	B	0.55	0.53	0.52	0.53	0.52	0.51
	A	0.56	0.54	0.53	0.55	0.53	0.52
(d) 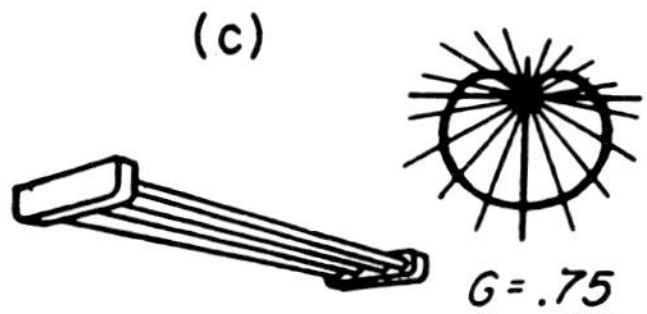 G=.75 A=.65 B=.55; 10/70	J	0.32	0.27	0.23	0.32	0.26	0.23
	I	0.40	0.35	0.31	0.39	0.34	0.30
	H	0.44	0.39	0.36	0.43	0.39	0.35
	G	0.48	0.43	0.40	0.46	0.42	0.39
	F	0.52	0.47	0.43	0.50	0.46	0.42
	E	0.57	0.52	0.46	0.55	0.51	0.47
	D	0.62	0.56	0.52	0.59	0.55	0.51
	C	0.65	0.59	0.54	0.62	0.57	0.54
	B	0.69	0.63	0.59	0.65	0.61	0.58
	A	0.71	0.66	0.62	0.67	0.63	0.60

* Sylvania Electric Products, Inc.

† G = good; A = average; B = bad maintenance factor.

Table 16-7. Coefficients of Utilization* (Continued)

Ceiling reflection factor, per cent		75			50		
Wall reflection factor, per cent		50	30	10	50	30	10
Cold-cathode fixture types†	Room index	Coefficients of utilization					
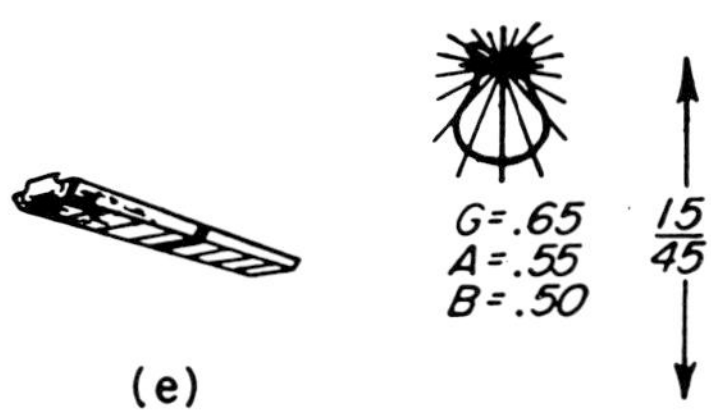	J	0.25	0.23	0.20	0.24	0.22	0.21
	I	0.32	0.28	0.27	0.29	0.27	0.26
	H	0.34	0.32	0.30	0.32	0.31	0.29
	G	0.37	0.35	0.33	0.35	0.33	0.31
	F	0.39	0.37	0.35	0.36	0.35	0.33
	E	0.42	0.40	0.38	0.39	0.38	0.36
	D	0.45	0.42	0.40	0.42	0.40	0.38
	C	0.46	0.44	0.41	0.43	0.41	0.40
	B	0.48	0.46	0.44	0.45	0.43	0.42
(e)	A	0.49	0.47	0.45	0.46	0.44	0.42
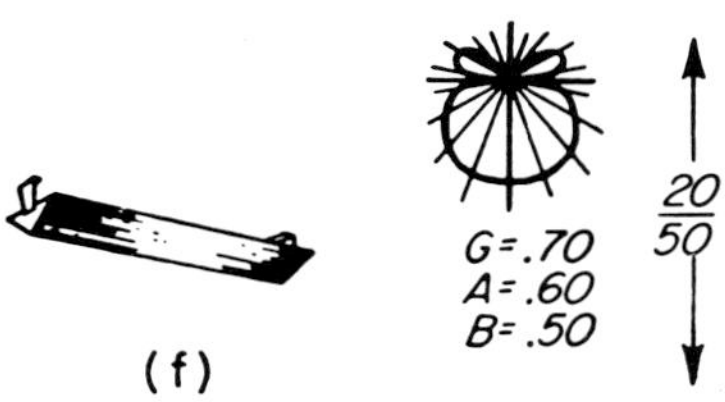	J	0.24	0.20	0.17	0.23	0.19	0.16
	I	0.30	0.26	0.23	0.28	0.24	0.22
	H	0.33	0.29	0.27	0.31	0.28	0.26
	G	0.36	0.32	0.29	0.34	0.31	0.28
	F	0.39	0.35	0.32	0.36	0.33	0.31
	E	0.42	0.39	0.36	0.40	0.37	0.35
	D	0.46	0.43	0.40	0.43	0.40	0.38
	C	0.48	0.45	0.42	0.44	0.42	0.40
	B	0.51	0.48	0.45	0.47	0.45	0.43
(f)	A	0.52	0.50	0.47	0.49	0.46	0.45
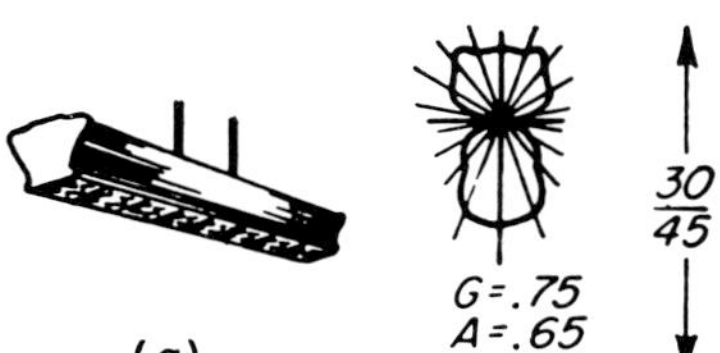	J	0.23	0.19	0.16	0.21	0.17	0.15
	I	0.28	0.24	0.22	0.26	0.22	0.20
	H	0.31	0.27	0.25	0.29	0.25	0.23
	G	0.35	0.31	0.28	0.31	0.28	0.26
	F	0.38	0.34	0.30	0.34	0.30	0.28
	E	0.42	0.38	0.34	0.37	0.34	0.31
	D	0.45	0.41	0.37	0.40	0.37	0.34
(g)	C	0.48	0.43	0.40	0.42	0.39	0.36
	B	0.51	0.47	0.44	0.45	0.41	0.39
	A	0.53	0.49	0.46	0.47	0.43	0.41
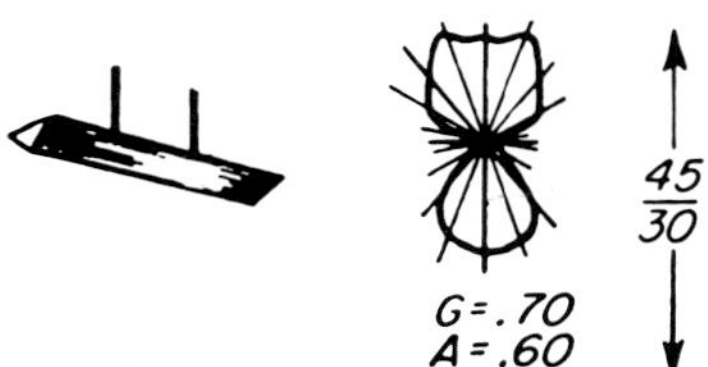	J	0.26	0.21	0.18	0.22	0.19	0.16
	I	0.31	0.26	0.24	0.27	0.24	0.22
	H	0.34	0.30	0.28	0.30	0.27	0.25
	G	0.38	0.34	0.31	0.34	0.30	0.28
	F	0.41	0.37	0.33	0.36	0.33	0.30
	E	0.45	0.41	0.38	0.40	0.36	0.34
	D	0.49	0.45	0.42	0.42	0.41	0.37
(h)	C	0.51	0.48	0.44	0.44	0.44	0.39
	B	0.55	0.51	0.49	0.47	0.45	0.43
	A	0.57	0.53	0.51	0.49	4.46	0.44

* Sylvania Electric Products, Inc.
† G = good; A = average; B = bad maintenance factor.

Table 16-7. Coefficients of Utilization* (Continued)

Ceiling reflection factor, per cent		75			50		
Wall reflection factor, per cent		50	30	10	50	30	10
Cold-cathode fixture types†	Room index	Coefficients of utilization					
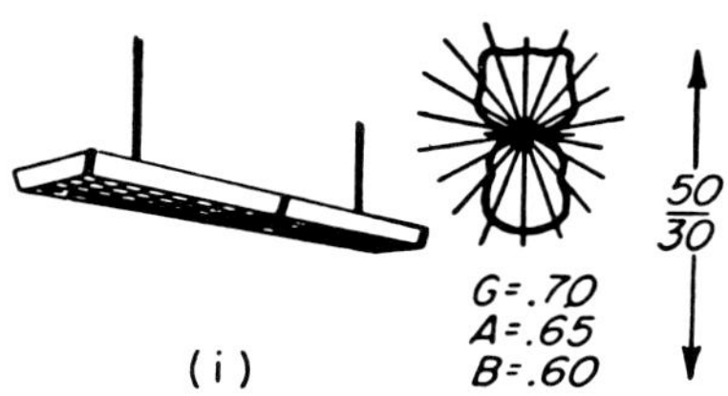 (i)	J	0.26	0.23	0.20	0.23	0.21	0.19
	I	0.31	0.28	0.27	0.28	0.26	0.24
	H	0.35	0.32	0.30	0.31	0.28	0.27
	G	0.38	0.35	0.33	0.34	0.31	0.30
	F	0.41	0.38	0.35	0.36	0.34	0.32
	E	0.44	0.42	0.39	0.39	0.37	0.35
	D	0.48	0.45	0.42	0.42	0.39	0.38
	C	0.50	0.49	0.44	0.43	0.41	0.39
	B	0.53	0.50	0.48	0.46	0.43	0.42
	A	0.54	0.52	0.50	0.47	0.45	0.43
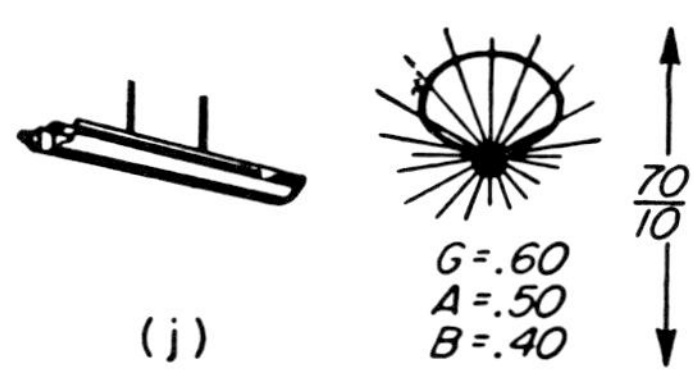 (j)	J	0.18	0.14	0.12	0.14	0.11	0.09
	I	0.22	0.19	0.17	0.17	0.15	0.13
	H	0.26	0.22	0.19	0.20	0.17	0.15
	G	0.29	0.25	0.22	0.22	0.19	0.17
	F	0.32	0.28	0.25	0.24	0.21	0.19
	E	0.35	0.32	0.29	0.27	0.24	0.21
	D	0.39	0.35	0.32	0.29	0.26	0.24
	C	0.42	0.38	0.35	0.31	0.28	0.27
	B	0.46	0.42	0.39	0.34	0.31	0.29
	A	0.48	0.44	0.42	0.36	0.33	0.31
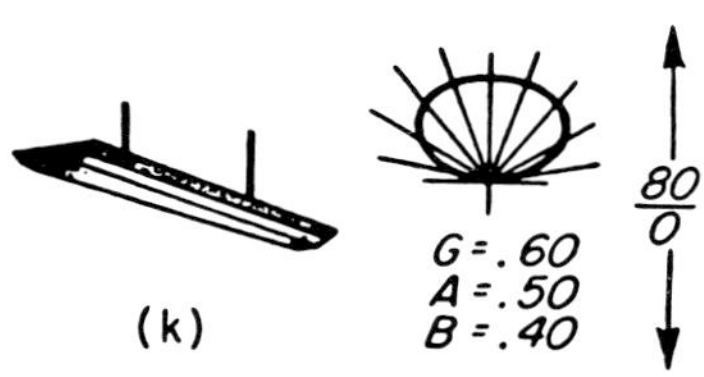 (k)	J	0.16	0.12	0.11	0.10	0.09	0.07
	I	0.20	0.16	0.14	0.13	0.11	0.10
	H	0.23	0.20	0.17	0.15	0.13	0.11
	G	0.27	0.23	0.20	0.18	0.15	0.14
	F	0.29	0.25	0.22	0.20	0.17	0.15
	E	0.33	0.29	0.26	0.22	0.19	0.18
	D	0.36	0.32	0.30	0.24	0.22	0.20
	C	0.39	0.35	0.32	0.26	0.23	0.22
	B	0.43	0.40	0.37	0.28	0.26	0.25
	A	0.45	0.42	0.39	0.30	0.28	0.26

* Sylvania Electric Products, Inc.

† G = good; A = average; B = bad maintenance factor.

Table 16-8. Room Index

Ceiling height, feet, for semi-indirect and indirect lighting		9 and 9½	10–11½	12–13½	14–16½	17–20	21–24	25–30	31–36	37–50		
Mounting height above floor, feet, for direct and semidirect lighting		7 and 7½	8 and 8½	9 and 9½	10–11½	12–13½	14–16½	17–20	21–24	25–30	31–36	37–50
Room width, feet	**Room length, feet**	Room index										
9 (8½–9)	8–10	H	I	J	J							
	10–14	H	I	I	J							
	14–20	G	H	I	J	J						
	20–30	G	G	H	I	J	J					
	30–42	F	G	H	I	J	J	J				
	42 and up	E	F	G	H	I	J	J				
10 (9½–10½)	10–14	G	H	I	J	J						
	14–20	G	H	I	J	J	J					
	20–30	F	G	H	I	J	J					
	30–42	F	G	G	H	I	J	J				
	42–60	E	F	G	H	I	J	J				
	60 and up	E	F	F	H	H	I	J				
12 (11–12½)	10–14	G	H	I	I	J	J					
	14–20	F	G	H	I	J	J					
	20–30	F	G	G	H	I	J	J				
	30–42	E	F	G	H	I	J	J				
	42–60	E	F	F	G	H	I	J				
	60 and up	E	E	F	G	H	I	J				
14 (13–15½)	14–20	F	G	H	H	I	J	J				
	20–30	E	F	G	H	I	J	J				
	30–42	E	F	F	G	H	I	J	J			
	42–60	E	E	F	F	H	I	J	J	J		
	60–90	D	E	E	F	G	H	J	J	J		
	90 and up	D	E	E	F	F	G	I	J	J		
17 (16–18½)	14–20	E	F	G	H	I	J	J				
	20–30	E	F	F	G	H	I	J				
	30–42	D	E	F	G	H	H	J	J	J		
	42–60	D	E	E	F	G	G	I	J	J	J	
	60–110	D	E	E	F	G	G	I	J	J	J	
	110 and up	C	D	E	E	F	G	H	I	J	J	
20 (19–21½)	20–30	D	E	F	G	H	I	J	J			
	30–42	D	E	E	F	G	H	I	J	J		
	42–60	D	D	E	E	F	G	I	J	J	J	
	60–90	C	D	E	E	F	G	H	J	J	J	
	90–140	C	D	D	E	F	F	H	I	I	J	J
	140 and up	C	D	D	E	F	F	H	H	I	J	J

Table 16-8. Room Index (Continued)

Ceiling height, feet, for semi-indirect and indirect lighting		9 and 9½	10–11½	12–13½	14–16½	17–20	21–24	25–30	31–36	37–50		
Mounting height above floor, feet, for direct and semidirect lighting		**7 and 7½**	**8 and 8½**	**9 and 9½**	**10–11½**	**12–13½**	**14–16½**	**17–20**	**21–24**	**25–30**	**31–36**	**37–50**
Room width, feet	**Room length, feet**	**Room index**										
24 (22–26)	20–30	D	E	E	F	G	H	I	J	J		
	30–42	C	D	E	F	G	G	I	J	J		
	42–60	C	D	D	E	F	G	H	I	J	J	
	60–90	C	D	D	E	F	F	H	I	J	J	J
	90–140	C	C	D	E	E	F	G	H	I	J	J
	140 and up	C	C	D	E	E	F	G	H	I	I	J
30 (27–33)	30–42	C	D	D	E	F	G	H	I	J	J	
	42–60	C	C	D	D	F	F	H	H	I	J	
	60–90	B	C	C	D	E	F	G	H	I	J	J
	90–140	B	C	C	D	E	E	F	G	H	I	J
	140–180	B	C	C	D	E	E	F	G	H	I	J
	180 and up	B	C	C	D	E	E	F	G	H	I	J
36 (34–39)	30–42	B	C	D	E	F	F	H	I	I	J	
	42–60	B	C	C	D	E	F	G	H	I	J	J
	60–90	A	C	C	C	E	E	F	H	H	J	J
	90–140	A	B	C	C	D	E	F	G	H	I	J
	140–200	A	B	C	C	D	E	F	F	G	H	I
	200 and up	A	B	C	C	D	E	F	F	G	H	I
42 (40–45)	42–60	A	B	C	C	E	F	G	H	I	I	J
	60–90	A	B	B	C	D	E	F	G	H	I	J
	90–140	A	B	B	C	D	D	E	F	G	H	J
	140–200	A	A	B	C	D	D	E	F	G	H	I
	200 and up	A	A	B	C	D	D	E	F	F	G	I
50 (46–55)	42–60	A	A	B	C	D	E	F	G	H	I	J
	60–90	A	A	B	C	C	D	F	F	G	H	J
	90–140	A	A	A	C	C	D	E	F	F	G	I
	140–200	A	A	A	C	C	D	E	E	F	G	I
	200 and up	A	A	A	C	C	D	E	E	F	G	H
60 (56–67)	60–90	A	A	A	B	C	D	E	F	G	H	I
	90–140	A	A	A	B	C	C	D	E	F	G	H
	140–200	A	A	A	B	C	C	D	E	E	F	H
	200 and up	A	A	A	B	C	C	D	E	E	F	H
75 (68–90)	60–90	A	A	A	A	B	C	D	E	F	G	I
	90–140	A	A	A	A	B	C	D	E	F	F	H
	140–200	A	A	A	A	B	B	C	D	E	F	G
	200 and up	A	A	A	A	B	B	C	D	E	F	G
90 or more	60–90	A	A	A	A	B	C	D	E	F	G	H
	90–140	A	A	A	A	A	B	C	D	E	F	G
	140–200	A	A	A	A	A	B	C	D	E	F	G
	200 and up	A	A	A	A	A	B	C	C	D	E	F

APPENDIX I

Luminous Tube Applications

The luminous tube has always had the potentials for wide ranging application for illuminated communications, directional information service, artistic enjoyment, architectural treatments, animated impact and for special light effects. But for most of the past half century, the medium has been used almost exclusively for sign illumination, ranging from simple identification statements to very sophisticated advertising marvels.

Neon lost its grasp as the major sign illumination source in the 1950s as the popularity of backlighted translucent acrylics proliferated and, for a long time during this period, sign manufacturers worked industriously to combine neon and plastic jointly in their design creations. But because of manufacturing economics and the then, new and expanding design flexibilities of plastics, the luminous tube demand failed. Tubing became relegated to a secondary role, but since the early '70s it has begun to enjoy a rekindled interest, not only for signage but for many other light expressions, including the art medium, architectural functions and interior design uses.

On the following pages is an illustrative selection of current day applications that depict tubing in an array of uses from functional to purely aesthetic dimensions.

Credits: Page 294, Fremont Street, Las Vegas—Signs of the Times, Cincinnati

Page 295, Holiday Inn sign—Signs of the Times, Cincinnati
Northcross Mall—Omniplan, Dallas

Page 296, Herschede store identity—Signs of the Times, Cincinnati
Pepsi Cola spectacular—Signs of the Times, Cincinnati

Page 297, Pedestrian sign—Lynn's Studio, Sioux Falls, S.D.
Traffic sign—Signs of the Times, Cincinnati
Parking lot sign—Signs of the Times, Cincinnati
Weather forecaster—Mercantile National Bank, Dallas

Page 298, Environmental exhibit—Signs of the Times, Cincinnati
Interior Eagle design—Harry Hake & Partners, Inc., Cincinnati
Bank of Vail sign—Signs of the Times, Cincinnati

Page 299, Nikon window display—Carl R. Leonard, Willoughby's, New York City
Fashion window—Nick Ewing, Liberty House, San Francisco

Page 300, "Walk on Neon"—Gilbert Associates, New York City
Neonized flower—Signs of the Times, Cincinnati
"American Dream"—Gilbert Associates, New York City

"Glitter Gulch" (Fremont Street in Las Vegas) is a neon spectacle.

Identity and ornamentation in neon and bulbs combination.

Logo and identity in exposed tubing.

Individual plastic-faced letters backlighted by tubing.

Animation achieved with flashing decorative tubing.

Traffic control signage, back-lighted (top, left) and exposed tubing.

Color-coded tubing provides weather forecasts.

Changing messages provide parking information.

Decorative tubing highlights exhibit message.

Supergraphic statement as a powerful interior design element.

Silhouette effect created by tubing behind letters.

Neon employed in a store promotion.

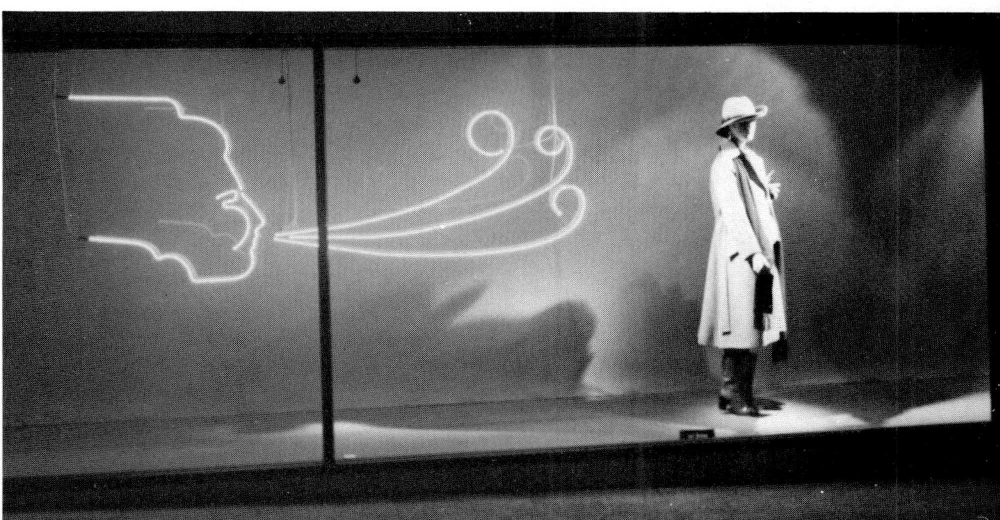

Luminous tubing creates visual impact for department store.

"Walk on Neon" by Stephen Antonakos

Neon sculptures create excitement and dimension as an art medium.

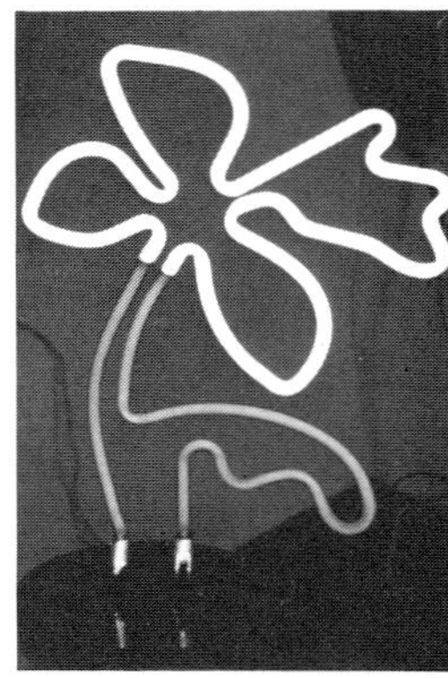

"American Dream" by Larry Rivers and Rudi Stern

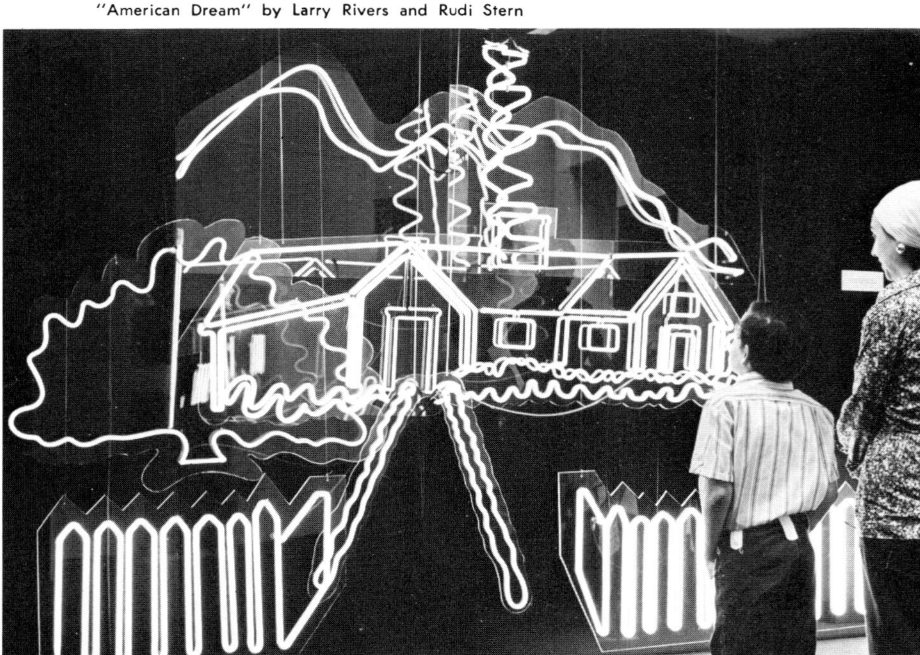

APPENDIX II

Let There Be Neon Tomorrow

Since 1915 the craft of neon has been exclusively in the domain of the electric signmakers. While George Claude had imagined many other applications for the process, he had synthesized that advertising gave commercial viability to the craft. Of course, the craft's technology and techniques developed as they were called into play by the realities of this market. Neon's graphic potential was well exploited in the Times Square of the 1930s. It was America's national piazza and the electric signmakers fused art and technology in boldly imaginative ways. The exuberance and vitality of America were reflected in her neon images.

The anonymous creaters of the neonized Ben Hur chariot races and the Wrigley's aquarium were, for me, artists of the first rank even if their skills remain mysterious alchemy even to this day. Neon provided popular electric folk art. It was the stage on which the American Dream took place. An early surplus of electrical energy was the reason that America's utilities promoted its use. Lindbergh traveled the country upon his return from Paris as a paid public relations agent for Claude Neon. He attributed his safe landing at Le Bourget to neon lights on the runway of the airfield and he lectured his audiences on the virtues of the imported medium. Neon was "refined" and "elegant" and an "important new art form." Early Claude Neon brochures said that it lent "dignity and sophistication" to promotion and displays.

The potentials of neon are great and without known boundaries. New directions in sculpture, architecture, scenic and interior design and environmental art are suggested by its inherent flexibility and kineticism. But for the moment there will be more interest in it because it is a very efficient source of light.

Architects, in my experience, fall into three categories in terms of their relationship to the medium: (1) Those who think of neon as the last pink "piazza" sign they have seen and who would never use it with all its supposed garish connotations of Las Vegas. These architects tend to want to know nothing about how neon works or how the tubing is made. For them it is a closed case. And an ugly one. (2) Those who take an intellectual and often rhapsodic view of the medium. They tend to like its nostalgic implications as well. Robert Venturi's "Revisiting Las Vegas" is a bit like the European discussion clubs that meet to delve into the existentialism of Batman. Taking essentially popular forms of communication and making them esoteric is an academic's luxury. (3) Architects who know nothing about luminous tubing but who are open to seeing in it various visual possibilities. They need to learn about it from first hand experience which the economy has not exactly been providing in recent times. Since successful applications lead to others, time with this group is well spent.

I believe that the future of neon and architecture looks promising because of the medium's efficiency, but it will be its graphic three-dimensionality that will ultimately give it popularity. The future of neon in architecture looks most promising for the following reasons:

(1) Neon is extremely efficient in terms of consumption of energy.

(2) As well as excellent illumination, it has graphic potential which incandescents and fluorescents cannot match. Neon is a fluid, three-dimensional, kinetic line. It is drawing with light in space. It is a shadowless source of light which can be molded into any existing architectural configuration or it can create its own spatial relationships.

(3) Neon lasts a very long time. I have old skeletons in my gallery which were pumped in 1945 and which are still intense. Re-pumping and re-gassing is a simple process which involves changing the electrodes. It takes approximately fifteen minutes to repump a unit. A transformer can work ten or fifteen years before it needs replacement. Both of these costs are minimal compared with weekly or monthly bulb changing. Neon is as permanent a light source as has ever been devised.

(4) The color possibilities of neon are great although tinted tubing is no longer made in this country. There are perhaps forty color variations that are possible in the juxtaposition of gas with tinted or coated glass. Architects generally are surprised at the range of color possibilities with neon and argon, and more of them would use the medium if they were aware of this range.

(5) Neon can be dimmed as easily as incandescents. Thus it is as capable of control as any other light source. It can be used very effectively with down lighting and controlled spots.

The potentials of neon as a sculptural medium are equally unexplored. There are at present approximately seventy sculptors worldwide using the medium as a primary means of expression. Of these seventy, perhaps ten bend their own glass, and the rest rely on the expertise of a local sign shop. Upon the level of communication between the designer and the glassbender depends the accuracy of a concept's realization. Obviously, the more the designer knows about the process, the greater the likelihood that the finished work resembles the original concept. So, too, the more open the glassbender is to the aesthetics of the designer, the more will his proficiency and craft lend themselves to the execution. There has to be a meeting between the two. If a glassbender uses nipples instead of tubulated electrodes, the result can be damaging for close-hand viewing. On a basic level, the glassbender has to adjust to the reality that the work, unlike outdoor signs, will not be seen from a distance of fifty feet, but might be viewed close-up and *in the round*. Mounting techniques that work well with an outdoor sign (metal boxes, pyrex housings, mounting posts, copper wire, electrical tape, corks, etc.) might not be the best solution to sculptural aesthetics.

Again, the designer has to know about the process to suggest worthwhile alternative methods to the glassbender. Often the best communication takes place when there is an interchange of information and when the glassbender learns about methods that also have a relevance for signage. The signman has to be open to new craft directions. He, too, has to expand his knowledge and awareness just as the designer/sculptor has to learn about the craft's traditions.

When America's electric sign companies turned to fluorescent lamps and plastic after World War II, many of the skills related to neon and outdoor signage in general deteriorated. Design suffered too as skills of glassbending and animation were lost. Glassbenders were often used for other work. There never has been a wholehearted apprenticeship program to train young glassbenders and few, if any, young people were trained in the uses of neon. There are many large electric sign companies in this country today which have no employees capable of designing or working with the medium. Sign salesmen sold plastic as the "new look" and cast neon as the "old look" with the result that layers of dust built up in many glass shops. Since today there may be as few as 200 glassbenders left in this country and since their average age might be above 50 one can say that the craft is in jeopardy. To discuss future uses of neon might be entirely hypothetical if there are no skilled craftsmen and craftswomen able to continue the technology. The most inventive uses of the medium will remain unknown unless the craft continues. At present there are five places in this country where neon is taught:

(1) Corky Marcheschi at the Minneapolis College of Art and Design teaches a course in kinetic sculpture which involves neon. He has set up a glassbending and bombarding facility for this purpose.

(2) Joe Reese teaches a neon course at the San Francisco Museum School which involves some glassbending.

(3) James Carpenter and Dale Gilhouley teach glass blowing at the Rhode Island School of Design and neon figures in some of their most innovative collaborative sculpture. Neon is not part of the curriculum but an element in their own work.

(4) Jerry Noe at the University of North Carolina in Chapel Hill teaches neon as part of his sculpture classes and he has set up a shop.

(5) Let There Be Neon in New York City has conducted workshops since 1973. Apart from sculptors, painters, and graphic artists, students have included photographers, architects, interior designers, a New York City fireman, an airline pilot, dentists, and an investment banker among many others. Eleven stations (each

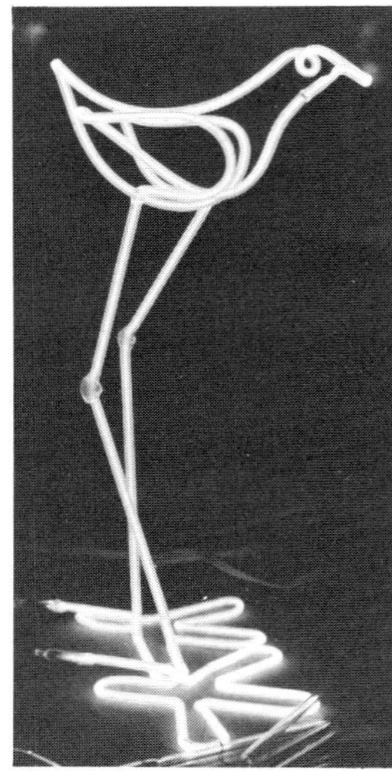

Clockwise from top left: (1) Neon nostalgia (photo by Standford Golob); (2) Freestanding Argon Bird by student John Tanaka; (3) Tom Scarff creates neon environments in natural settings; (4) a neonized facade in Teheran; (5) "Cyclone" by Ron Steinhilber.

with three fires) are provided. The Workshop consists of individual instruction geared to a particular student's needs and capabilities. Many continue glassbending after their five or ten week course is completed. Casper Ingui, a glassbender with more than 25 years experience, teaches this aspect of the curriculum. Michael Hauenstein and I discuss the history, craft, and three dimensional applications of the medium. Visiting guest lecturers have included sculptors, signmakers, and neon craftsmen skilled in various aspects of the process. Many of the completed student works have been shown at the Let There Be Neon Gallery.

It is crucial to recognize the medium as a handcraft. The public and architects in particular have to be made aware of its existence as a learned skill. Behind the deceptive ease with which a good glassbender works is a highly trained sensitivity to the materials. Like any handcraft it has a history and a tradition. Recognizing these is a first and important step in dealing with its future.

—Rudi Stern, New York, N.Y.

RUDI STERN: Rudi Stern has been working with neon as an art medium for more than ten years. His first neon sculpture (in collaboration with Jackie Cassen) was made for the Architectural League of New York in 1967. He opened Let There Be Neon in 1972 as a gallery devoted to neon art, both past and present. The Workshop began in 1973. Apart from his work with tubing, he has done extensive work with light projection (opera, ballet, theater, environments, etc.), kinetic sculpture and video. He has received numerous grants from the Rockefeller Foundation, the New York State Council of the Arts, the National Endowment on the Arts and the John D. Rockefeller III Fund. He has recently completed a book on neon art which will be published in 1977 by Overlook Press. Stern's biography is included in the 1977 edition of Who's Who in the East.

Index

T